To Sylvia,
and to our daughters,
Sandra and Zoe

CAUSATION

A REALIST APPROACH

Michael Tooley

CLARENDON PRESS · OXFORD

*This book has been printed digitally and produced in a standard design
in order to ensure its continuing availability*

OXFORD
UNIVERSITY PRESS

Great Clarendon Street, Oxford OX2 6DP
Oxford University Press is a department of the University of Oxford.
It furthers the University's objective of excellence in research, scholarship,
and education by publishing worldwide in

Oxford New York

Athens Auckland Bangkok Bogotá Buenos Aires Cape Town
Chennai Dar es Salaam Delhi Florence Hong Kong Istanbul Karachi
Kolkata Kuala Lumpur Madrid Melbourne Mexico City Mumbai Nairobi
Paris São Paulo Singapore Taipei Tokyo Toronto Warsaw

with associated companies in Berlin Ibadan

Oxford is a registered trade mark of Oxford University Press
in the UK and in certain other countries

Published in the United States
by Oxford University Press Inc., New York

ISBN 0-19-824962-4

PREFACE AND ACKNOWLEDGEMENTS

It was a graduate seminar taught by Carl G. Hempel that first led me to think in a serious way about causation, and about laws of nature. The views towards which I have subsequently gravitated are metaphysically rather remote from those that I held then. However I hope that, between the metaphysical lines, something of Hempel's influence is visible.

A number of philosophers have given me very detailed comments on different parts of various drafts of this book. In this regard, I am greatly indebted to David Lewis, Jack Smart, Bas van Fraassen, and Evan Fales.

The comments, criticisms, and suggestions that I have received from a number of other philosophers, either in discussions, or in response to my earlier papers on laws of nature and causation, have helped me greatly. I should therefore like to express my appreciation to John Burbidge, Leonard Carrier, Mendel Cohen, John Collins, Ed Erwin, Peter Forrest, Lannie Goldman, Danny Goldstick, Stephen Hetherington, Herbert Hochberg, Frank Jackson, Ramon Lemos, Clifton McIntosh, Mohan Matthen, Howard Pospesel, Richard Routley, Richard Sharvy, Francis Snare, David Stove, Chris Swoyer, and Peter Valentine.

By far my greatest debt, however, is to David Armstrong. The detailed comments that he offered on every part of this book, together with the many discussions that we have had, both on laws of nature and on causation, have significantly affected my thinking on most issues in this area. This will be especially clear in Part II, much of which obviously reflects our continuing dialogue on the nature of laws.

I am also very indebted to the two anonymous readers for Oxford University Press. The book is much stronger as a result of their incisive comments and helpful suggestions.

The research for this book, and the writing of it, took place at four universities. The initial draft was completed while I was a Research Fellow, and later a Senior Research Fellow, in the Philosophy Department of the Research School of Social Sciences

at the Australian National University. Major revisions were
carried out while I was at the University of Miami, and further
revisions were completed here at the University of Western
Australia. The extensive final revisions were then undertaken
while I was on study leave at Trent University. I am very grateful
to all four universities for their support, and for providing a highly
congenial atmosphere in which to work. In addition, I am
especially indebted to the Law School of the University of Western
Australia for providing me with access to their word-processing
equipment.

Parts of this book are based upon two previously published
pieces. Chapters 2 and 3 contain some material from my article,
'The Nature of Laws', which appeared in the *Canadian Journal of
Philosophy*, 7/4, 1977, pp. 667–98, while Chapter 6 is a slightly
revised and condensed version of my paper, 'Laws and Causal
Relations', which was published in *Midwest Studies in Philosophy*,
9, Minneapolis (Copyright 1984 by the University of Minnesota)—
though I now take a rather different view concerning the
conclusions to be drawn from the arguments developed there. I am
grateful to the publishers in question for permission to use this
material.

Finally, I need to mention some of the conventions that are used
throughout the book to facilitate concise reference to properties,
states of affairs, and predicates. Suppose that 'F' is a predicate,
and that there is some property P such that 'F' is true of something
if and only if that entity has property P. How can one refer to
property P in such a way that the connection with the predicate 'F'
is indicated? In informal contexts, one could speak of the property
associated with predicate 'F', or, more briefly, of the property of
F-ness. In formal contexts, one could use a property abstraction
operator that maps predicates into their associated properties.
Such devices are, however, rather cumbersome. Indeed, in formal
contexts where proofs are being offered, the use of property
abstraction operators applied to predicates greatly increases the
amount of space needed for the statement of proofs.

The convention adopted here is as follows. If 'F', 'G', 'H' are
predicates, then rather than speaking of the properties F-ness,
G-ness, and H-ness, I shall speak simply of properties F, G, and
H. Similarly, in the relevant formal contexts, rather than using
an expression such as '$(\lambda x)\ (Fx) \to^* (\lambda x)\ (Gx)$' to express the

proposition that the property associated with the predicate '*F*' is related via the ancestral of direct causal necessitation to the property associated with the predicate '*G*', I shall write '*F*→* *G*'.

A similar problem arises in the case of states of affairs. One often wants to refer to the state of affairs by virtue of which some sentence '*Fa*' is true. In informal contexts, one could speak of the state of affairs associated with the sentence in question, while, in formal contexts, one could introduce a state of affairs abstraction operator mapping true sentences into the corresponding states of affairs. A much less cumbersome method, however, is to parallel the approach used in the case of properties, and to speak simply of the state of affairs *Fa*.

Finally, there is the case of predicates. As expressions referring to predicates will not be employed in more formal contexts, the need for a concise way of referring to predicates is less urgent. Nevertheless, it still seems desirable to avoid a proliferation of single quotes, and this can be done if one refers to predicates by a convention akin to that of display, rather than by means of quotation-names. Thus I shall generally speak of the predicates *F*, *G*, *H*, rather than of the predicates '*F*', '*G*', '*H*'. I shall, however, use quotation-names in the case of predicates in ordinary language.

Michael Tooley

Perth, Australia
March 1987

CONTENTS

LIST OF SYMBOLS

Introduction

This book is concerned with the nature of causation, and it represents an attempt to develop an account that falls within the empiricist tradition, broadly conceived, but which, at the same time, diverges radically from previous accounts in certain crucial respects. The book also contains an extended discussion of the nature of laws—a topic intimately related to that of causation.

The most important difference between the accounts of causation and of laws to be offered here, and traditional empiricist accounts, is that the latter are reductionist, in the sense that they entail that once all non-causal, non-nomological facts are fixed, it is thereby logically determined both what laws there are, and what states of affairs or events are causally related. Consequently the first part of the book deals with this central difference, and with the foundations of a realist approach. For, as we shall see, a realist account of causation and laws does involve some significant ontological commitment. In particular, it presupposes realism both with respect to universals and with respect to theoretical entities.

The second part of the book is concerned with the nature of laws. The view which is advanced is that laws are to be identified with certain contingent relations among universals. The case of strictly universal laws is discussed in Chapters 2 and 3, and the analysis is then extended to the case of probabilistic laws in Chapter 4.

The third, and central part of the book deals with causation. Chapter 5 briefly describes some of the more important issues that need to be considered in setting out an account of the nature of causation. In Chapter 6, the question of the supervenience of causal relations upon non-causal facts, plus causal laws, is discussed, and a number of arguments are offered against a supervenience view. The central problem of the direction of causation is then considered in Chapter 7. Current accounts are examined, in order, first, to bring out the difficulties confronting a reductionist approach, and second, to isolate some plausible constraints upon any satisfactory account.

A realist theory of the nature of causation is then set out, and defended, in Chapters 8 and 9. The basic idea is that causal laws and relations, rather than being analysable in terms of probabilistic ones, as some philosophers have suggested, are to be viewed instead as theoretical relations that *underlie and explain* probabilities. It is then shown how this idea can be captured and developed in a systematic way, and it is argued that the resulting theoretical-relation approach to causation avoids the objections to which reductionist accounts are typically exposed, and that it does so without making causal laws and causal relations epistemologically inaccessible.

PART I

Philosophical Background

— 1 —

Foundations of a Realist Approach

Many accounts of the nature of causation and of laws have been advanced, but none has elicited anything remotely approaching general acceptance. Perhaps this testifies only to the extreme difficulty of providing satisfactory analyses of fundamental concepts. The approach to be taken here, however, suggests that the apparent intractability of the problem may also have another source. For a conspicuous feature of the present accounts of causation and of laws is that they do involve significant ontological commitment. Thus it will be argued, first, that laws are to be identified with certain contingent, irreducible, theoretically specified relations among universals, and second, that causal relations between states of affairs are also theoretically specified relations, and ones that are not reducible to non-causal properties and relations. The present accounts therefore presuppose, first, that there are universals; second, that it is possible for them to stand in contingent, irreducible relations; and third, that realism with regard to theoretical entities is a defensible position. To the extent, then, that the accounts to be advanced here turn out to be plausible, and to the extent that they avoid difficulties that commonly confront alternative approaches, there is reason for thinking that questions of ontology may very well play a much more crucial role in connection with the problems of the nature of causation and of laws than is so for many philosophical problems, and thus that some of the difficulties encountered by other attempts to provide analyses of causal and nomological concepts may reflect over-restrictive ontological assumptions. In later chapters I shall try to offer reasons for thinking that this is in fact the case.

In the present chapter I shall begin by discussing three important, underlying issues. First, the existence of universals. Second, the intelligibility of a realist view of theoretical terms. Third, the question of whether causal and nomological terms really stand in need of analysis. I shall then conclude by considering how the approaches to causation and laws to be developed here diverge from the dominant empiricist tradition which derives from David Hume, and the extent to which, given that divergence, the present approach can be viewed as still falling within the general empiricist tradition.

I.I THE EXISTENCE OF UNIVERSALS

The meaning of saying that there are universals is perhaps most simply explained by means of some claims concerning properties and relations. It is important to note, however, that the terms 'property' and 'relation' are used in two very different senses. On the one hand, there is the usage—undoubtedly the dominant one among philosophical logicians—according to which these terms are employed to refer to intensional entities, so that properties are monadic *concepts*, and relations polyadic ones. On the other hand, there is the usage favoured by metaphysicians, according to which the terms 'property' and 'relation' are used to refer to non-intensional entities, to objective aspects of things in the world. It is the latter usage which I shall follow here.

The claim that there are universals may now be explained as involving acceptance of at least the following three claims. First, individuals do have properties, and stand in relations, and having properties and standing in relations cannot be explained in terms of class membership. For properties differ from classes in that P and Q may be distinct properties, even though everything that has property P also has property Q, and vice versa. Second, if two individuals have some property in common, they must be genuinely identical in some respect, where this cannot be a matter simply of falling under the same concept, or under the same predicate. Thus something which is blue, and something which is red, necessarily fall under some of the same concepts—such as the concept of not being green—but this provides no reason at all for saying that there is some respect in which they are genuinely the same, and thus no reason for holding that they share some property. Finally, to claim that there are universals is to advance a certain account of what it is for individuals to be identical in a certain respect. It is to say that their being identical in a certain respect is a matter of there being some entity—a universal—which is wholly present in both individuals.

If laws are to be identified with certain relations among universals, then one is certainly committed to the existence of universals. But one needs to ask whether the account of the nature of laws to be advanced might not be reformulated in a more nominalistic fashion, so that the reference to universals is eliminated. For many statements that are apparently about universals can be replaced by other statements which do not explicitly refer to universals, and without any loss of content—beyond the implication that there are universals. Thus, for example, the statement 'The universals roughness and roundness are, as a matter of fact, always co-instantiated' can without serious loss be replaced by the statement 'All and only rough things are round'—thus eliminating the reference to a relation between universals. So perhaps the view that laws are relations between universals can also be given an ontologically less robust formulation.

This possibility certainly needs to be considered. Other things being equal, the fewer ontic commitments the better. In Part II, however, I believe that it will become clear that such a reformulation is not available in the case of the view that laws are relations among universals. For I shall attempt to show that truth conditions for nomological statements cannot be given in terms of states of affairs which involve only particulars having properties and standing in relations: one needs to refer to *irreducible* relations among universals, where an irreducible relation among universals is one that cannot be analysed in terms of properties of, and relations among, particulars. Moreover, as we shall also see, the reference to universals enables one to explain some features of nomological statements that are very difficult to handle on other accounts, such as the intensionality of nomological statements, and the fact that such statements can be highly confirmed, on the basis of a small number of properly selected observations, even in an infinite universe.

At the same time, it seems clear that the commitment to universals will render the account somewhat less appealing for many philosophers. For while it is probably true that relatively few contemporary philosophers accept nominalism—understood as 'the doctrine that whatever exists is a particular, and nothing but a particular'[1]—this widespread rejection of nominalism is rarely reflected in the ways in which philosophical problems are tackled. Philosophers seem, on the whole, disinclined to make any use of universals in grappling with philosophical problems in areas outside of metaphysics itself. This reluctance is perhaps most vivid in contemporary philosophy of language. Thus, in offering accounts of key semantical notions such as truth, reference, and meaning, quantification over properties and relations, realistically conceived, is almost universally eschewed; while reference to particulars of types whose nature, at least upon reflection, is far from unproblematic—such as sets, and possible worlds, and even Meinongian objects—is widely embraced.

Given that an acceptance of universals is crucial for the present enterprise, it may be helpful to consider, at least briefly, some of the more important reasons for thinking that nominalism is not a tenable view.[2] First, there is the 'One over Many' argument. This rests upon the observation that two particulars can be, and often are, the same in some respect. Two pens may be of precisely the same shade of red. Two fundamental particles may be of precisely the same mass. The issue then arises as to what account is to be offered of the sameness. The realist with respect to universals holds that it is a matter of one and the same thing—a universal— being wholly present in both particulars. Nominalists reject this account. But the question is whether a nominalist can in fact provide an alternative account which is satisfactory, and which does not turn out to involve reference to universals.

Some nominalists seem to feel that there is no serious problem here. Thus Michael Devitt, for example, has suggested, in response to the above line of argument, that a sentence such as '*a* and *b* have the same property (are of the same type), *F*-ness' need merely be construed as equivalent to the sentence '*a* and *b* are both *F*'. And so he concludes that 'there is no problem about identities in nature beyond a trivial one of paraphrase'.[3]

However, this response reflects a failure to grasp the basis of the present argument. In particular, the One over Many argument

---

[1] Armstrong, *Universals and Scientific Realism*, i: *Nominalism and Realism*, Cambridge, 1978, p. 138.

[2] For a very thorough statement of the case against nominalism, see Armstrong, ibid., chaps. 2–5.

[3] Devitt, ' "Ostrich Nominalism" or "Mirage Realism"?', *Pacific Philosophical Quarterly*, 61, 1981, pp. 433–9. See p. 435.

does not rest upon the fact that two things may fall under the same concept, or under the same predicate. Realists would grant—indeed insist—that the fact that two objects fall under the same concept—such as the concept of being either a stone or an angel, or the concept of not being green—provides no reason for holding that the objects must be genuinely identical in some respect. But at the same time they would hold that some objects are genuinely identical in one or more respects: electrons, for example, are identical with respect to mass and with respect to charge. It is the latter sort of fact that is the basis of the One over Many argument, and that the realist is claiming can only be satisfactorily explained by postulating universals.

In failing to distinguish between these two very different sorts of facts—that some objects fall under the same concept or under the same predicate, and that some objects are genuinely identical in some respects—Devitt is in effect assuming that the realist maintains that there must be universals corresponding to *all* general predicates. It is somewhat surprising that Devitt makes this mistake, since his article is directed specifically against Armstrong's discussion in *Universals and Scientific Realism*, and Armstrong emphasizes very strongly that not all general predicates are associated with corresponding universals. In chapter 14 of volume ii, for example, Armstrong argues that neither negative universals nor disjunctive universals exist—that is, if *P* and *Q* are predicates associated with distinct, non-overlapping universals, then there cannot be any universals corresponding to the following predicates: not-*P*, not-*Q*, *P* or *Q*.

This point about universals and predicates explains why most versions of nominalism do not have the resources needed for an explanation of the facts in question. For just as there is no one-to-one correspondence between universals and predicates, so neither is there any one-to-one correspondence between universals and concepts, or between universals and classes. The upshot is that predicate nominalism, concept nominalism, and class nominalism all appear unable to offer any response to the One over Many argument.

The only version of nominalism which really seems to address itself to this argument, and tries to provide a solution, is resemblance nominalism. But resemblance nominalism involves serious difficulties of its own. Of these, the most familiar is

10   *Philosophical Background*

probably the regress argument formulated by Bertrand Russell, which is directed to showing that an appeal to the relation of resemblance can at best reduce the number of universals that one needs to postulate, since one is always left with at least one universal—namely, the relation of resemblance.[4]

A second classical argument for the existence of universals is the argument from meaning: universals must be postulated to serve as the meanings of general predicates. In its traditional form, this argument is doubly unsatisfactory. First, it confuses universals with concepts. Second, it assumes that for every predicate there is a corresponding universal. Nevertheless, I believe that there is the core of a sound argument here, which can be extracted once the two erroneous suppositions are eliminated. What one is left with, it would seem, is the problem of explaining why a given general predicate applies to some particulars but not to others. The realist has an answer: whether or not a predicate applies to a given particular is determined by the properties of the particular, and the relations into which it enters. In contrast, it does not seem that the nominalist has a satisfactory account to offer—though here too the resemblance nominalist at least addresses the problem.

A third argument in support of the existence of universals was developed by Arthur Pap, and has recently been strengthened by Frank Jackson.[5] Consider the statement:

Red resembles orange more than blue.

A realist has no difficulty offering an analysis of this statement, namely:

Any red-making property (i.e. any property by virtue of which something is red) resembles any orange-making property more than any blue-making property.

What analysis might a nominalist offer: One try is this:

Anything red resembles anything orange more than anything blue.

[4] Russell, *The Problems of Philosophy*, London, 1912, chap. 9. Philosophers are divided on the soundness of Russell's argument. For a discussion of the argument, together with a formulation of other objections to resemblance nominalism, see Armstrong, op. cit., chap. 5.
[5] Pap, 'Nominalism, Empiricism and Univerals: I', *Philosophical Quarterly*, 9, 1959, pp. 330–340; Jackson, 'Statements About Universals', *Mind*, 86, 1977, pp. 427–9.

But as Pap points out, this statement is not even true, since 'x may resemble z more than y *in other respects* though x is red and y orange and z blue'.[6] So it cannot provide an adequate analysis of the original statement.

The natural response to this difficulty is to replace the predicate 'resembles' by the predicate 'colour-resembles', giving one the following analysis:

> Anything red colour-resembles anything orange more than anything blue.

This proposal presupposes that it is legitimate to treat 'colour-resembles' as a primitive predicate. But even if this assumption is granted for the sake of discussion, it is easy to show that the proposed analysis does not provide a translation of the original statement. For consider the possible world in which all and only red things are triangular, all and only orange things are sweet, and all and only blue things are square. In that world it would be true that anything triangular colour-resembles anything sweet more than anything square. Yet one would not want to say that, in that world, triangularity resembles sweetness more than squareness.[7] At this point, the nominalist may note that the original statement 'Red resembles orange more than blue' is true not only in the actual world, but in all possible worlds—or at least in all possible worlds in which there are red, orange, and blue things. This suggests the following analysis:

> Necessarily, anything red colour-resembles anything orange more than anything blue.

The use of the modal operator will, of course, make this proposal unappetizing to many nominalists. But even if that issue is waived, it is clear that the revised analysis will not do. For statements of colour comparisons that are merely contingently true—such as 'The colour of ripe tomatoes resembles the colour associated with girl babies more than the colour associated with boy babies'—will turn out to be false on the analysis proposed.[8]

A nominalist who is a realist with regard to possible worlds can, however, improve on the above, and can do so without holding that 'colour-resembles' is a primitive predicate. For if one is

---

[6] Pap, op. cit., p. 334.   [7] Jackson, op. cit., p. 428.
[8] Ibid., p. 429.

prepared to take a realist view of unactualized entities, one can hold that the statement that red resembles orange more than blue is to be analysed as:

> Any red thing resembles some orange thing more than it resembles any blue thing

—where the quantifiers in the latter statement range over everything in every possible world.[9]

That the above statement is analytically equivalent to the statement

> Red resembles orange more than blue

assuming that there are unactualized entities, certainly seems true. None the less, I think it is implausible to regard it as providing an analysis. For it is natural to say that if there are unactualized entities, with the consequence that it is true that any red thing resembles some orange thing more than it resembles any blue thing, then there is an explanation of the latter fact. The explanation is that, while for any red thing there will be a blue thing in some possible world that differs from the red thing only with respect to colour, there will also be, in some possible world, an orange thing that differs from the red thing only with respect to colour, and then, *since* red resembles orange more than blue, the red thing will resemble the orange thing more than it will the blue thing. In short, it is because red resembles orange more than blue that it is true—assuming the existence of unactualized entities— that any red thing resembles some orange thing more than it resembles any blue thing, and so the former statement cannot be analysed in terms of the latter.

The above line of argument appears to be a strong objection to virtually all forms of nominalism. The one version where there may be some ground for doubt is particularism. This is the view that the properties and relations of ordinary particulars are themselves particulars on the same level. Ordinary, or 'thick' particulars are thus, on this view, bundles of 'thin' particulars, which have been variously referred to as 'abstract particulars', 'particularized qualities', 'unit properties', 'tropes', 'cases', and

---

[9] This proposal is a slight variant on that suggested by Lewis, in 'New Work for a Theory of Universals', *Australasian Journal of Philosophy*, 61/4, 1983, pp. 343–77. See n. 10, p. 349.

'property-instances'.[10] For given this version of nominalism, it might be suggested that the original statement can be analysed as follows:

> Any particular redness resembles any particular orangeness more than it resembles any particular blueness.

The idea is that by formulating the analysis in terms of 'thin' particulars, rather than in terms of ordinary particulars, one can avoid the difficulty that arises from the fact that red things may resemble blue things more than orange things because of other properties. But if this difficulty is to be avoided, it must be the case that particularized qualities do not themselves possess any properties, other than necessary ones, and I cannot see what reason there is for thinking that this is so. Consider, for example, a particular case of redness. That instance of redness will have a certain shape. In the case of an object that is red all over, the shape of the redness will be the shape of the object. As a consequence, one might try to maintain that it is the ordinary particular that has the shape, not the redness. But if only part of the ordinary particular is red, it seems difficult to avoid attributing a shape to the redness itself. In short, there would appear to be reason for holding that particularized qualities, if they exist, sometimes possess contingent properties, and this means that the above analysis cannot be satisfactory; the world might very well contain some particular redness that, by virtue of its shape, more closely resembled some particular blueness than it did any particular orangeness. It is doubtful, therefore, that particularism can provide an answer to the third argument against nominalism.

I.2 A REALIST CONSTRUAL OF THEORETICAL TERMS

There is a second ontological issue that has a crucial bearing upon the problem of providing truth conditions for causal and nomological statements. This is the question of the possibility of a realist construal of theoretical terms. For it will be argued, in Chapters 2 and 4 in the case of nomological statements, and in Chapters 7 and 8 for causal statements, that the relevant truth-makers are

---

[10] Armstrong, op. cit., i, pp. 78–9.

theoretical states of affairs, and that no reductionist account of those states can provide satisfactory truth conditions.

It is quite essential for our purposes, then, to have at hand an intelligible realist construal of theoretical terms. In this section I shall describe such an account.

Theoretical statements, by definition, cannot be analysed in purely observational terms. From this, many have inferred that if one confines oneself to analysis in the strict sense, theoretical statements can be analysed only in terms of other theoretical statements: there is no possibility of translating such statements into ones that are free of all theoretical vocabulary. But it is clear that this does not follow, since the class of statements that are free of theoretical vocabulary does not coincide with the class of observation statements. For example, the statement 'This table has parts too small to be observed', although it contains no theoretical vocabulary, does refer to things that are unobservable, and so is not a pure observation statement. This situation can arise because, in addition to observational vocabulary and theoretical vocabulary, one also has logical and quasi-logical vocabulary—including expressions such as 'part', 'property', 'event', 'state', 'particular', and so on—and statements containing such vocabulary, together with observational vocabulary, can refer to unobservable states of affairs.

This suggests the possibility, which I believe to be correct, that theoretical statements, though not analysable in terms of observation statements, are analysable in terms of statements that contain nothing beyond observational, logical, and quasi-logical vocabulary. The most natural and straightforward way of carrying out such an analysis is that set out and defended in detail by David Lewis in his article, 'How to Define Theoretical Terms',[11] and the approach that I shall adopt here is a slightly modified version of it—the modification reflecting the fact that I am operating within a different metaphysical framework from that of Lewis.

Lewis's method of defining theoretical terms is a development of the treatment suggested by F. P. Ramsey in his essay, 'Theories'.[12] It may be helpful, therefore, to begin by considering Ramsey's

[11] Lewis, 'How to Define Theoretical Terms', *Journal of Philosophy*, 67, 1970, pp. 427–46.
[12] Ramsey, 'Theories', *The Foundations of Mathematics*, ed. Braithwaite, Paterson, NJ, 1960, pp. 212–36.

approach. Let T be any theory. If T contains any singular theoretical terms, eliminate them by paraphrase. Then replace all theoretical predicates and functors by expressions involving names of corresponding properties and relations, so that, for example, an expression like '. . . is a neutrino' is replaced by an expression such as '. . . has the property of neutrino-hood'. The result of these changes can then be represented by

$$T(P_1, P_2, \ldots P_i, \ldots P_n)$$

where each P_i is the theoretical name of some property or relation. All such theoretical names are then replaced by distinct variables, ranging over properties and relations, and the corresponding existential quantifiers prefixed to the formula. The resulting sentence

$$(\exists U_1)\,(\exists U_2) \ldots (\exists U_i) \ldots (\exists U_n)\, T\,(U_1, U_2, \ldots U_i, \ldots U_n)$$

is what is known as a Ramsey sentence for theory T. Ramsey's suggestion, then, is that such an existentially quantified statement, which is free of all theoretical terms, can be used in place of the original theory T.

It is important to appreciate the point of this proposal. To do so we need to distinguish between realist and anti-realist intepretations of theories. The distinction is based upon the answers given to the following two questions:

(1) Do the theoretical terms of theory T have a denotation?
(2) Is it possible to analyse or reduce all statements about such entities to statements about the observable properties and relations of observable objects?

To adopt a realist interpretation of theory T is to give an affirmative answer to (1) and a negative answer to (2), while, on an anti-realist construal of theory T, either the answer to (1) is negative or the answer to (2) is affirmative.

The two ways of rejecting a realist interpretation of a theory give rise to different versions of anti-realism: (*a*) the reductionist view, (*b*) the instrumentalist view. The reductionist agrees with the realist that theoretical terms do have a denotation, but he holds that the entities thus referred to are nothing more than possible elaborate 'constructs' out of observable states of affairs. The instrumentalist takes a more radical view of theories. He holds that theoretical terms have no denotation at all. The point of a theory, according to the instrumentalist, is not to say something about the world. A theory is simply a formal device that maps statements about what is observable into other statements about what is observable, and that provides some sort of systematization of truths about the observable world.

How are theories to be construed, if they are approached along the lines suggested by Ramsey? It appears that the choice between a realist and an anti-realist interpretation remains open. For on the one hand, it seems clear that Ramsey himself opted for an anti-realist interpretation of theories. He says, for example, that the meaning of a theoretical statement, *p*, relative to some set, *S*, of statements, can be identified with the difference between the *observational* consequences of *S*, together with *p*, and of *S* alone. He also says that relations between theories—such as equivalence, incompatibility, and so on—can be understood in terms of the relations of those theories to observable facts.[13]

But on the other hand, a Ramsey-type approach coheres at least as well with a realist interpretation of theories. Indeed, it offers the realist a plausible answer to a very pressing question. For one of the central problems for the realist interpretation of theories is how it is possible to attach meanings to theoretical terms if they are construed as referring to what is unobservable. A Ramsey approach to theories provides the realist with the following answer: theoretical terms are intelligible if the theories containing them are taken as asserting nothing beyond what is asserted by the corresponding existentially quantified statements from which the theoretical terms have been removed.

How satisfactory an answer is this to the question of the meaning of theoretical terms, realistically construed? If the verifiability criterion of meaning were correct, it would surely be of no value. The shift from the original theory to the corresponding Ramsey sentence does nothing to illuminate how statements purporting to refer to unobservable entities can possible be verified. However, most philosophers feel that there are good reasons for rejecting the verifiability theory of meaning. One

---

[13] See Ramsey, 'Theories', p. 231 for the former, and pp. 232–3 for the latter. Scheffler, in the discussion of the interpretation of theories in *The Anatomy of Inquiry*, New York, 1963, classifies Ramsey's approach as a type of 'eliminative fictionalism'. See pp. 203–22.

which is especially significant in the present context is that the verifiability theory of meaning rules out a realist interpretation of theoretical terms. The reason is this: let T be some theory which purports to assert the existence of unobservable entities, and let T^* be the statement 'All observable events are just as they would be if T were true.' According to the verifiability criterion of meaning, there cannot be any difference between T and T^* with respect to cognitive content. But T^* does not assert the existence of anything unobservable. So T cannot assert the existence of anything unobservable either.[14]

Let us adopt, then, the alternative view that the meaning of a statement is adequately explained if one can show how it can be translated into a statement containing only antecedently understood vocabulary. Given this perspective, Ramsey's approach seems very promising. There are, however, some objections that need to be considered.

First, consider the following two, rather simple, theories:

$$T_1 \qquad\qquad T_2$$
$$(x)(Px) \qquad (x)(\sim Px)$$
$$(x)(Px \supset Ox) \qquad (x)(\sim Px \supset Ox)$$

where P is a theoretical term, and O observational. Corresponding to these two theories we have the Ramsey sentences:

$$R(T_1)\colon (\exists U)[(x)(Ux) \ \& \ (x)(Ux \supset Ox)]$$
$$R(T_2)\colon (\exists V)[(x)(\sim Vx) \ \& \ (x)(\sim Vx \supset Ox)].$$

Commenting on these two theories, and their respective Ramsey sentences, Jane English says: 'Here two theories that share a body of observational consequences, yet contradict each other in their theoretical consequences, are found to have Ramsey sentences that are perfectly compatible—indeed, their Ramsey sentences are equivalent.'[15]

English has advanced two claims here. The one is that

[14] This argument presupposes that the verifiability criterion of meaning has been formulated in such a way as to provide a criterion of *sameness* of cognitive content. For evidence that this is how it was traditionally understood, together with reasons why it ought to be thus formulated, see Salmon 'Verifiability and Logic', *Mind, Matter, and Method*, ed. Feyerabend and Maxwell, Minneapolis, Minn., 1966, pp. 354-76.

[15] English, 'Underdetermination: Craig and Ramsey', *Journal of Philosophy*, 70, 1973, pp. 453-63. See p. 458.

incompatible theories may have logically equivalent Ramsey sentences. She believes that this first contention can be established by the following simple argument. Take the Ramsey sentence $R(T_1)$, and replace both occurrences of Ux with $\sim\sim Ux$, giving:

$$R(T_1)^*: (\exists U)[(x)(\sim\sim Ux) \ \& \ (x)(\sim\sim Ux \supset Ox)].$$

But if there is some U such that $[(x)(\sim\sim Ux) \ \& \ (\sim\sim Ux \supset Ox)]$ is true, then there must be some V, namely, whatever corresponds to $\sim U$, such that $[(x)(\sim Vx) \ \& \ (x)(\sim Vx \supset Ox)]$ is true. So $R(T_1)^*$ entails $R(T_2)$, and since the former is entailed by $R(T_1)$, we have the result that $R(T_1)$ entails $R(T_2)$. In similar fashion, by, so to speak, replacing $\sim V$ with U, we can derive $R(T_1)$ from $R(T_2)$. So $R(T_1)$ and $R(T_2)$ are logically equivalent.

If one holds, both that there are universals, and that the existential quantifiers at the beginning of a Ramsey sentence should be taken as ranging over universals, rather than over properties and relations in some broader sense, then a quick answer to this first argument is immediately evident. For if the existential quantifiers range over universals, and if, as suggested above, there are no negative universals, then $\sim U$ cannot range over universals, and so it cannot be legitimate to replace $\sim U$ with V, since the latter will range over universals. English's argument for the claim that the Ramsey sentences are equivalent therefore either involves the uncritical assumption that there are universals corresponding to all coherent predicates, or presupposes some broader notion of properties and relations.

This response leaves untouched, however, the weaker contention that English advances above—the claim, namely, that the Ramsey sentences of the two incompatible theories, regardless of whether they are equivalent, are at least compatible. And if this latter claim can be sustained, it is surely a very damaging objection to a Ramsey approach.

But can this objection be sustained? The difficulty of doing so emerges if one asks what determines the meaning (or meanings) of the occurrences of P found in T_1 and T_2. Apparently there are two alternatives. The one is that the meaning of an occurrence of P is implicitly defined by the theory in which it occurs. The other is that the meaning is somehow independently specified.

The first alternative cannot be employed in the context of the present objection, for if the meaning of an occurrence of P is implicitly defined by the theory in question then one has a very good reason for holding that occurrences of P in T_1 do not have the same meaning as occurrences of P in T_2. They are occurrences of different predicates. Thus theories T_1 and T_2 are only apparently incompatible.

Suppose, then, that the meaning of predicate P is independently specified in some manner. The statement, p, by means of which this is done, will then have to be conjoined to T_1 and T_2 to produce expanded theories, T_1^* and T_2^*, which incorporate everything that is relevant to the meaning of the terms contained in T_1 and T_2. The question is then whether the Ramsey sentences of the expanded theories will be compatible. The answer is that they will not be. For $R(T_1^*)$ will say that everything possesses that unique property, whatever it may be, that satisfies condition p, whereas $R(T_2^*)$ will say that nothing possesses that unique property, whatever it may be, that satisfies condition p.

In short, theories T_1 and T_2, while appearing incompatible, are not genuinely incompatible unless it is the same predicate P that occurs in T_1 and T_2, and once one ensures that this is so, one is confronted with modified theories, T_1^* and T_2^*, whose Ramsey sentences are also incompatible. So English's first objection cannot be sustained.

Another objection that English offers to the view that a theory can be equated with its Ramsey sentence turns upon the related issue of whether there can be consistent theories that are incompatible, but that have all the same observational consequences. English claims that this would not be possible if a theory were just its Ramsey sentence.

If this objection is sound, the idea of using a Ramsey approach to make sense of a realist approach to theories must be abandoned. For one cannot be a realist without holding that there can be consistent incompatible theories that have all the same observational consequences.

English's argument may be put as follows. Let S and T be two mutually incompatible, consistent theories that have precisely the same observational consequences. Let $R(S)$ and $R(T)$ be corresponding Ramsey sentences. Now it may very well be the case that S and T share some theoretical vocabulary. If so, systematically modify T by replacing all occurrences of those theoretical terms that also occur in S by occurrences of new theoretical terms, and do this in such a way as to produce a new theory, T^*, that is isomorphic with T. Clearly, T^* will have precisely the same observational consequences as T, and the Ramsey sentence of theory T—$R(T)$—will also be a Ramsey sentence for theory T^*.

The crucial question now is this: Is it possible for the Ramsey sentences $R(S)$ and $R(T)$ to be incompatible? Let us assume that they are. Then, in view of the fact that any theory entails all of its Ramsey sentences, we have that S entails $R(S)$ and T^* entails $R(T)$, and thus that S and T^* must be incompatible if $R(S)$ and $R(T)$ are incompatible. So there must be a proof, from the hypothesis that S, of not-T^*. But now one can appeal to Craig's interpolation theorem, which states that if there is a proof in first order predicate logic of C, on the basis of A, then there exists some formula B, whose non-logical vocabulary is restricted to what is common to A and C, and which is such that B is derivable from A, and C from B. Applying this to the present case leads to the conclusion that there must be some formula, M, which is entailed by S, and which in turn entails not-T^*, and whose non-logical vocabulary is restricted to the non-logical vocabulary shared by S and not-T^*. Since T^* was defined in such a way that it shares no theoretical terms with S, the shared non-logical vocabulary must be purely observational. Therefore there is an observational statement, namely M, which is entailed by S, and whose negation is entailed by T^*. Since T and T^* have precisely the same observational consequences, and since S and T are consistent theories, it follows that there is an observational statement, namely M, which is entailed by S but not by T. The assumption that the Ramsey sentences, $R(S)$ and $R(T)$, are incompatible has therefore been shown to be incompatible with the hypothesis that S and T are consistent theories with the same observational consequences. Therefore, if theories were just their Ramsey sentences, it would be impossible for there to be mutually incompatible, consistent theories with precisely the same observational consequences.[16]

This is an interesting argument, but it is unsound. The defect will emerge if we consider the following two simple theories:

S: $(x)(x$ has property $P)$
$(x)(x$ has property $P \supset x$ has property $O)$

[16] Ibid., pp. 459–60.

$(P \neq O)$

$(U)[(x)(x$ has property $U \supset x$ has property $O) \supset ((U = P)$

v $(U = O))]$

T: $(x)(x$ has property $Q)$

$(x)(x$ has property $Q \supset x$ has property $O)$

$(Q \neq O)$

$(\exists U)[(x)(x$ has property $U \supset x$ has property $O)$ & $(U \neq$

$Q)$ & $(U \neq O)]$

where O is an observational term, and P and Q are theoretical terms. Now both S and T are consistent theories; they have precisely the same observational consequences; and they are mutually incompatible. But in the case of these two theories, the Ramsey sentences $R(S)$ and $R(T)$ are also incompatible, since $R(S)$ entails that there is only one property, distinct from property O, whose presence ensures the presence of O, while $R(T)$ entails that there are at least two such properties.

Where, then, is the error in English's argument? The answer is that her argument is sound up to and including the point where she claims that, if $R(S)$ and $R(T)$ are incompatible, there must be some statement, M, which is entailed by S, and which entails not-T^*, and whose non-logical vocabulary is purely observational. The error lies in inferring that M must therefore be an observation statement. For as is illustrated by statements such as 'Tables have parts that are too small to be observed' and 'There are at least two properties, distinct from O whose presence is sufficient to ensure the presence of O', the mere absence of non-observational, descriptive vocabulary does not suffice to make a statement an observation statement.

To sum up, the situation with regard to incompatible theories and their corresponding Ramsey sentences is this: there are two natural ways of attempting to construct consistent theories that have the same observational consequences, but that appear to be incompatible. The first involves shared theoretical vocabulary. When this approach is taken, it is easy to construct theories that appear to be incompatible, but whose Ramsey sentences are not. The problem is that one cannot provide any grounds for the claim that the theoretical terms that are 'shared' by the two theories have the same meaning in the different theories, without introducing statements which force one to expand the theories into ones that do have incompatible Ramsey sentences.

The second approach involves theories which contain quantifiers ranging over properties and/or relations. This enables one to construct theories that are genuinely incompatible, even though they do not differ with respect to their observational consequences. But such theories constitute no objection to a Ramsey approach.

A third objection to Ramsey's approach is this. If one replaces a theory by a corresponding Ramsey sentence, it is true that one is no longer confronted with the question of the meaning of certain theoretical terms. But for some purposes, it would seem that one needs to be able to offer some account of the meaning of theoretical terms, rather than merely avoiding their use. It does seem to be an intelligible question to ask, for example, how the concept of mass in Newtonian physics compares with the concept of mass in relativistic physics. To answer this sort of question, one needs analyses of the two concepts; the Ramsey approach as it stands does not provide any analyses of theoretical terms, it merely shows how one can dispense with them.

The Ramsey approach is, however, easily modified to cope with this request for analyses of theoretical terms. The development needed is carefully set out by Lewis in his paper 'How to Define Theoretical Terms' and it is to his discussion that I shall now turn. Again, let T be any theory, and eliminate singular theoretical terms, and theoretical predicates and functors, in favour of complex predicates involving names of the associated properties and relations. Represent this result by:

$$T(P_1, P_2, \ldots P_i, \ldots P_n)$$

where $P_1, P_2, \ldots P_i, \ldots P_n$ are all the theoretical names in question. Now replace each of these by distinct variable letters, giving:

$$T(U_1, U_2, \ldots U_i, \ldots U_n)$$

One can now offer the following account of the meaning of the theoretical name, P_i:

x has property P_i

means the same as

There is a unique *n*-tuple of properties and/or relations, $(K_1, K_2, \ldots K_i, \ldots, K_n)$, which satisfies the open formula $T(U_1, U_2, \ldots U_i, \ldots U_n)$, and x has K_i.[17]

One can then use these definitions to eliminate the relevant theoretical terms from any statement containing them, including the original theory *T*. The result will be a statement which entails the Ramsey sentence but is slightly stronger. For the statement in question, unlike the Ramsey statement, will assert not merely that there is some *n*-tuple of universals which satisfies the relevant open formula, but that there is exactly one such *n*-tuple.

Let us now consider a final objection which can be directed both against Ramsey's original approach and against Lewis's supplemented version of it. The objection focuses upon the rather attenuated theory, *T*, which consists of the single statement:

$(x)(x$ has property $P \supset x$ has property $O)$

where *P* is theoretical, and *O* observational. This theory has the peculiarity that its Ramsey sentence

$(\exists U)(x)(x$ has property $U \supset x$ has property $O)$

is a logical truth. Since the original theory is not a logical truth, this example seems to show that a theory cannot just be its Ramsey sentence.[18]

What is the situation, given Lewis's approach? To answer this question, one needs to consider how the term 'property' is to be interpreted. Lewis himself favours a liberal interpretation of the term 'property' that makes it the case that there is a property for every predicate, and many more besides: a property can be thought of as any mapping from possible worlds into sets of entities, however disparate the entities in question may be. Given

[17] Lewis, op. cit., p. 438, mentions this type of account, but prefers instead to define theoretical terms by means of identity statements. This involves explaining the meaning of theoretical terms by means of statements such as 'Property P_i is identical with the *i*th member of the unique *n*-tuple of properties and/or relations which satisfies the open formula, "$T(U_1, U_2, \ldots U_i, \ldots U_n)$"'. Lewis points out that for these latter definitions to work out properly, one needs to adopt a treatment of denotationless names along the lines suggested by Scott in 'Existence and Description in Formal Logic', *Bertrand Russell: Philosopher of the Century*, ed. Schoenman, London, 1967.

[18] This objection is advanced by Scheffler, op. cit., pp. 218 ff.

this interpretation it will be true, for any property *O*—even the impossible property—that there is more than one property *P*, such that if something possesses property *P* then it does, as a matter of fact, possess property *O*. But if there is no unique property, then theory *T* does not succeed in assigning meaning to the term '*P*'.

The question then becomes whether this consequence constitutes an objection to Lewis's account. It might be thought that it does, on the ground that, however uninteresting theory *T* may be, it is at least intelligible, and possibly true. However it seems to me that the idea that theory *T* is intelligible, and possible true, arises because one takes it as implicitly part of the theory, first, that *P* and *O* are distinct properties, and secondly, that the term 'property' is used in a narrower sense than that favoured by Lewis. Suppose, in particular, that one restricts the term 'property' to universals. Then the theory that one is really considering could be put explicitly as follows:

$$T^*: (x)(x \text{ has universal } P \supset x \text{ has universal } O) \ \& \ (P \neq O)$$

Since property *P* may now very well be unique, Lewis's approach is perfectly compatible with the claim that T^* could be true. The present objection can, therefore, be set aside.

One final point calls for comment. Suppose that one is more sceptical than Lewis, either concerning the existence of genuine possible worlds, or concerning the existence of sets. Then properties, in Lewis's broad sense, may turn out to be mind-dependent objects, and it might well seem undesirable to carry out one's analysis of the meaning of theoretical terms in a way that involves reference to such entities. One might then prefer to adopt, as I shall, a mild variation of Lewis's approach which quantifies over universals, rather than properties in the sense of mappings from possible worlds into sets of entities. In some cases this will mean, of course, that further paraphrase of the original theory may be necessary, before the Ramsey/Lewis method can be applied: not all predicates have single universals associated with them, and those that do not will have to be eliminated in favour of predicates that do. But once this is done, the analysis proceeds as before.

In conclusion, then, I believe that there is good reason for holding that the realist about theoretical entities has available a satisfactory general account of the meaning of theoretical terms,

and one which is compatible with different views concerning what
sorts of things are ontologically basic.

1.3 CAUSAL AND NOMOLOGICAL CONCEPTS: THE NEED
FOR ANALYSIS

Many attempts have been made to offer analyses of fundamental
causal and nomological concepts. The difficulty in finding satisfactory accounts might lead one to ask what reason there is for
thinking that it is important to offer *analyses* of such concepts, or
even for thinking that it is possible. After all, at least with
reference to causal concepts, a strong case can be made for the
view that some such concepts are very familiar indeed. Language
abounds, for example, with transitive verbs—such as 'touch',
'move', 'lift', 'cut', 'see', 'hear', and so on—that are used to refer
to actions or occurrences that involve some causal relation, and an
understanding of many of these is acquired by children at a very
early age.[19] So why should it be necessary, or even possible, to
offer analyses of such concepts? May it not be perfectly reasonable
to treat such fundamental concepts as primitive and unanalysable?

What distinguishes terms that may be taken as ultimately
primitive, and terms that stand in need of analysis? A plausible
view, I suggest, is that primitive, non-logical vocabulary is
restricted to terms that satisfy the following three conditions:

1. Every primitive predicate must have a single universal
   associated with it, and every primitive singular term must
   refer to a single individual.
2. The universals associated with primitive predicates, and the
   individuals referred to by primitive singular terms, must be
   ones with which one is directly acquainted.
3. No primitive predicate can be such that elementary statements containing that predicate necessarily involve non-
   extensional contexts.

The idea underlying the first part of condition 1 is that, if a given
predicate sometimes applies to an individual by virtue of that
individual's having one property, and sometimes by virtue of an

---

[19] Anscombe, 'Causality and Determination', *Causation and Conditionals*, ed.
Sosa, Oxford, 1975, pp. 63–81. See pp. 68–9.

individual's having some other property, then it must surely be possible to make explicit the way in which the applicability of the predicate depends upon the various universals by analysing elementary statements involving that predicate in terms of statements containing predicates, each of which is associated with only one of the universals in question.

The second part of condition 1 is an analogous requirement in the case of singular terms, and a comparable rationale can be offered. If a singular term sometimes refers to one individual and sometimes to another, then it would seem that it must be possible to provide an analysis that makes it clear why the term in question sometimes refers to one individual and sometimes to another.

In the case of predicates that can be analysed, it is possible to understand their meaning without being acquainted with the property (or relation) that they function to attribute to individuals (or n-tuples of individuals). Similarly, one can understand a singular term if it is analysable, without being acquainted with the referent. But if the term in question is one that cannot be analysed, it would seem that there is no way that one can understand the term unless one is acquainted, in the case of singular terms, with the referent, and, in the case of predicates, with the property (or relation) associated with it. Thus we have condition 2.

The rationale underlying condition 3 might be put as follows. Some statements may be true by virtue of very complex states of affairs; but in the case of logically elementary statements that are free of quantifiers, and that contain only primitive terms, the truth-makers should be of a correspondingly simple sort; the state of affairs that makes such a basic statement true should be either a matter of an individual having a certain property, or of a number of individuals standing in a certain relation. Suppose, however, that F is a predicate whose occurrence in elementary statements involves non-extensional contexts, so that it is possible for Fa to be true, and Fb to be false, even though a is identical with b. Then it cannot be the fact that the individual picked out by the term a has the property associated with the predicate F which makes Fa true, for in that case Fb would be true as well. The truth of Fa must also be a function of the way in which the term a characterizes the individual in question. The statement therefore does not have the sort of simple truth-maker that is appropriate for logically elementary statements containing only primitive vocabulary. Terms that introduce non-extensional contexts cannot, for that reason, be treated as primitive.

If these requirements are justified, it seems clear that causal and nomological terms cannot be treated as primitive. In the case of nomological terms, there are at least two reasons why this is so. The first is that nomological terms introduce non-extensional contexts. Consider, for example, the expression 'it is a law that . . .' and assume that it is true both that

(1) It is a law that sugar dissolves in water, and that

(2) Water is the liquid most drunk by humans.

Then the predicates 'is in water' and 'is in the liquid most drunk by humans' would be coextensive, and the substitution of the latter for the former in sentence (1) would yield:

(3) It is a law that sugar dissolves when it is in the liquid most drunk by humans.

Sentence (3) may very well be false, however, even if both (1) and (2) are true. So the expression, 'it is a law that . . .', introduces non-extensional contexts, and thus, in view of condition 3, cannot be treated as a primitive expression.

The other objection to treating nomological expressions as primitive arises out of the second requirement. Let P be any predicate, and suppose that it is true that for any statement S, containing P, it is possible to construct a statement T satisfying the following conditions: (*a*) S entails T; (*b*) all of the vocabulary of T is primitive; (*c*) T does not contain P, nor any predicate that is analytically equivalent to P; (*d*) S and T are experientially equivalent, in the sense that it will make no difference to anyone's experience whether S is true, or T is true. When this situation obtains, it seems clear that P cannot be a primitive term. For if statements S and T are experientially equivalent in spite of the fact that S contains predicate P while T does not, there would seem to be only three possible explanations: (i) T contains some term synonymous with P; (ii) no property or relation with which one can be directly acquainted is associated with P; (iii) P can be analysed in terms of the vocabulary contained in T. Possibility (i) is precluded by hypothesis. But both possibilities (ii) and (iii) entail that P cannot be a primitive term: in the case of possibility (ii), that follows by virtue of the second requirement on primitive terms, whereas in the case of possibility (iii), the consequence is an immediate one. It follows, therefore, that whenever the above situation obtains, P must be analysable.

Let us now apply this to the case of nomological terms. The basic idea is simply that even if it turns out to be true that laws are not reducible to regularities, it will still be the case that there is no *observable* difference between a world that contains certain regularities, and a world that contains both regularities and underlying laws. This means that for any nomological statement there will be an experientally equivalent statement that refers only to the corresponding, Humean regularities, plus any other observational consequences of the nomological statement. For if S is a statement containing nomological terms, one can take T to be simply the conjunction of all of the statements that follow from S, and that are free of nomological terms. S and T will then be experientially equivalent, and will satisfy the other conditions as well. So it follows that the nomological terms in S cannot have associated with them any properties or relations with which one can be directly acquainted. Nomological terms cannot, therefore, be treated as primitive and unanalysable.

Precisely the same argument can be advanced in the case of causal terms. For here too, even if it turns out that some non-reductionist account of causation is correct, it will still be true that there is no *observable* difference between a world in which all of the non-causal facts are as they would be if states of affairs were causally related, and a world in which the states of affairs in question really are causally related. If S is any statement containing causal terms, one can take T to be simply the conjunction of all the statements that are entailed by S, and that are free of causal terms. S and T will be experientially equivalent, and will satisfy the other conditions. So there cannot be any properties or relations, with which one can be directly acquainted, that are associated with causal terms. Consequently, neither causal terms, nor nomological terms, can be treated as primitive, however familiar some of them may be. Analysis is required.

I.4 HUMEAN SUPERVENIENCE

Most contemporary accounts of causal and nomological concepts, although they diverge from Hume's views in various ways, are, at bottom, fundamentally Humean in inspiration. What I have in mind here can, perhaps, best be brought out by reference to two theses concerning the truth conditions of causal and nomological statements. I shall refer to these as *theses of Humean supervenience.*[20]

The first thesis of Humean supervenience concerns the truth conditions of nomological statements:

Thesis 1: The truth values of nomological statements are logically determined by the truth values of non-nomological statements about particulars.

This thesis can be expressed in other ways. Put in terms of possible worlds, the thesis would be that there cannot be two possible worlds that agree with respect to all non-nomological facts about particulars, but disagree with respect to what laws there are. The basic idea, then, is simply that once all non-nomological facts about particulars are fixed, so are all the laws.

This thesis marks one of the great gulfs between Humean views of nomological statements, and most non-Humean views. For a Humean, a law is just a cosmic regularity. Fix the non-nomological facts about particulars, and you have determined what cosmic regularities, and so what laws, there are. But a non-Humean, in contrast, thinks that there could in principle be cosmic regularities that were merely grand accidents—just cosmic coincidences. This in turn means, for most non-Humeans, that there could be distinct worlds that agreed with respect to all non-nomological facts about particulars, and so with respect to all cosmic regularities, but which did not agree with respect to all laws, since where there was a law in one world there might be merely a cosmic coincidence in the other.

The second thesis of Humean supervenience deals with the truth conditions of singular causal statements:

Thesis 2: The truth values of all singular causal statements are logically determined by the truth values of statements of causal laws, together with the truth values of non-causal statements about particulars.

---

[20] This expression was suggested by Lewis's discussion in 'A Subjectivist Guide to Objective Chance', *Studies in Inductive Logic and Probability*, ii, ed. Jeffrey, Berkeley and Los Angeles, Calif., 1980, pp. 263–93. See pp. 290–2.

This thesis asserts that once all non-causal facts about particulars, and all causal laws, are fixed, then it is logically settled which events, or states of affairs, are causally related. This claim would be accepted by all Humeans, but, in contrast to Thesis 1, it would also be accepted by most non-Humeans. Some, however, would reject it. In particular, if one adopts a singularist view of causation, and holds that it is causal relations between states of affairs, or events, that are primary, rather than causal laws, then one may want to reject Thesis 2. For if causal relations are primary, there could be a world devoid of causal laws, but containing causally related states of affairs. In such a world, it might be possible to have the following situation: (1) events C_1 and C_2 have precisely the same non-causal properties, both relational and non-relational; (2) E_1 and E_2 have precisely the same non-causal properties, both relational and non-relational; (3) the same non-causal relations hold between C_1 and E_1 as between C_2 and E_2; (4) although C_1 is the cause of E_1, C_2 is not the cause of E_2.

It should perhaps be emphasized that this possibility is not entailed by a singularist view of causation by itself. For it is possible to accept a singularist view, but to hold that causal relations are reducible to non-causal properties and relations, in which case Thesis 2 would be acceptable, although it would contain superfluous reference to causal laws. If, however, one both accepts a singularist account, and rejects a reductionist analysis, then it would seem that one must reject the second thesis of Humean supervenience.[21]

If one were asked to formulate a significant proposition concerning causation that would be very widely and firmly accepted among present-day philosophers, the claim that a singularist conception of causation must be rejected in favour of a supervenience view would probably be an excellent candidate. Nevertheless, we shall see, in Chapter 6, that any supervenience account is exposed to a number of rather serious objections. I shall argue, however, that the correct response to those objections is not to adopt a singularist approach. For there is a third view that can be taken, and which escapes both the objections that can be

---

[21] One philosopher who appears to adopt this sort of position is Anscombe, op. cit. In contrast, Ducasse, in his article, 'On the Nature and the Observability of the Causal Relation', ibid., pp. 114–25, appears to combine a singularist account with a reductionist analysis.

directed against a supervenience account, and the basic difficulty of intelligibility that arises from a singularist view. The two theses set out above capture, I suggest, ideas that lie at the heart of Humean approaches to causation, and to laws. They are also propositions that are only rarely challenged in present-day discussion. For this reason, I believe that one can say that almost all current discussions of the nature of laws, and particularly of causation, however much they may differ from Hume's views in various respects, are fundamentally Humean in inspiration.

1.5 SUMMARY AND OVERVIEW

To what extent do the accounts to be offered here of the nature of laws, and of causation, fall outside the dominant Humean tradition? The answer is that the divergence is mainly with respect to the issue of supervenience, but there the differences are very great. This is especially so with regard to the account of laws. For if the truth-makers for nomological statements are contingent, irreducible relations among universals, then it is impossible for laws to be supervenient upon non-nomological facts about particulars. So the first thesis of Humean supervenience must be abandoned.

In the case of causation, in contrast, the situation is different, since my account of the nature of causation does not entail that the thesis of Humean supervenience with regard to singular causal statements must be rejected. Nevertheless, there is still a serious divergence from the Humean tradition, in that my account is *neutral* on the issue of supervenience. Developed in one way, supervenience obtains. Developed in another, it does not. Humean accounts, on the other hand, are committed to supervenience.

Moreover, although the general approach to the analysis of causal concepts to be developed below can accommodate a supervenience view, it seems to me, as I mentioned above, that there are very strong independent arguments against the thesis that causal relations are supervenient upon causal laws, together with non-causal facts about particulars. So although it is not a consequence of the general approach that I favour, I do believe that there are good reasons for abandoning the second thesis of Humean supervenience, as well as the first.

In other respects, however, my accounts of the nature of causation, and of laws, do fall squarely within the main empiricist tradition, as they do involve the acceptance of a number of points upon which Hume, and those who have followed him, have insisted. In the first place, then, there is the issue of the status of statements expressing possible laws of nature, either causal or non-causal. On the account to be offered here such statements are, if true, only contingently so; they express neither logically necessary truths, nor synthetic a priori truths.

Secondly, there is the issue of the status of general principles such as the principle of universal causation, or the principle of sufficient reason. Here, too, I wish to maintain that such general principles are at best contingent truths; they can be neither logically necessary, nor synthetic a priori.

Thirdly, there is the question whether, even if statements expressing laws do not have the status of necessary truths, either logical or synthetic a priori, there may not be some a priori restrictions upon what sorts of events, or states of affairs, can be causally related to one another. I also wish to agree with Hume on this point: any type of contingent state of affairs involving particulars could in principle be causally related to any other.

Fourthly, there is the question of whether causal and nomological concepts can be treated as primitive. As I indicated in section 1.3, I believe that there is good reason for holding that such concepts cannot be treated as primitive. If they are to be admissible, an analysis must be forthcoming, and one which, moreover, satisfies constraints that have their roots in the general empiricist tradition.

To sum up: my basic methodological approach to the analysis of causal and nomological concepts is essentially that of the analytic and empiricist tradition going back to Hume. The analyses arrived at, however, diverge quite radically from that tradition, especially with regard to the issue of supervenience. The explanation of this rather sharp divergence is ontological in nature. For, as I have indicated above, it is my belief that one of the reasons that the problem of providing acceptable analyses of causal and nomological concepts has proved so intractable is that previous attempts to analyse these concepts have burdened themselves with either, or both, of the following unjustified assumptions:

(1) Either universals do not exist, or if they do, it is never necessary, in stating truth conditions for sentences, to refer to states of affairs into which universals enter *as individuals* having higher-order properties, or standing in higher-order relations to other universals.

(2) Theoretical terms either must be interpreted non-realistically, or else can be so intepreted without loss.

I have suggested that both assumptions deserve to be rejected, and in what follows I shall attempt to show that when they are rejected, the way is clear for the formulation of adequate empiricist accounts of causal and nomological concepts.

PART II

Laws

— 2 —
Laws: The Basic Approach

The next three chapters deal with the question of the nature of laws. In Chapters 2 and 3 the discussion will be confined to strictly universal, non-statistical laws. For since the case of probabilistic laws raises some additional issues, it seems best to tackle non-probabilistic laws first. If a satisfactory solution can be found, one can then see whether it is possible to extend it to the more complex case of probabilistic laws. This latter task will be taken up in Chapter 4, where I shall attempt to show that the analysis proposed for non-statistical laws can in fact be extended, in a natural manner, to the case of probabilistic laws.

The fundamental thesis to be defended is that it is possible to set out an acceptable, non-circular, extensional account of the truth conditions of nomological statements if and only if contingent, irreducible, theoretical relations among universals are taken as the truth-makers for such statements.

The idea that laws are relations among universals is not a new one. Indeed, as Armstrong points out in his book on the nature of laws, it may even be that something like this view is to be found in Plato's *Phaedo*.[1] However this idea never really took root, and it was only comparatively recently that this approach to laws was independently rediscovered, and then set out in a more explicit and sustained fashion, by Fred Dretske, David Armstrong, and myself.[2]

By far the most complete and systematic defence of the view that laws are relations among universals is found in Armstrong's *What Is a Law of Nature?* As many readers will be familiar with Armstrong's account, it may be helpful to indicate briefly, at this

[1] Armstrong, *What Is a Law of Nature?*, Cambridge, 1983, p. 85. Plato, *Phaedo*, sect. 102–7.
[2] Dretske, 'Laws of Nature', *Philosophy of Science*, 44, 1977, pp. 248–68. Armstrong, *Universals and Scientific Realism*, ii, Cambridge, 1978, chap. 24. Tooley, 'The Nature of Laws', *Canadian Journal of Philosophy*, 7/4, 1977, pp. 667–98.

point, the main ways in which the present account diverges from
that advanced by Armstrong. These differences will then be
discussed in more detail later.

There would appear to be three main differences between the
two accounts. The first concerns the possibility of laws that have,
as a matter of fact, no instances. Armstrong holds that all genuine
laws must be instantiated at some time and place, whereas I
maintain that this is not the case.

The difference needs, however, to be explained a little more
fully. At first glance, it might seem that Armstrong's view is open
to the rather plausible objection that something like Newton's
First Law of Motion could very well be a law even if there were
never a case where the net force acting upon some body was
equal to zero. But this particular type of case of a possibly
uninstantiated law is not, in fact, a serious objection to Armstrong's
theory, since, as he argues, one can use the term 'law' in an
extended sense that will include such cases, provided that one is
willing to analyse it in terms of statements about what laws (in the
strict sense) there would be, if certain conditions obtained. Thus,
for example, the claim that Newton's First Law of Motion might
be an uninstantiated law can be interpreted as expressing the
subjunctive conditional claim that if there were cases of bodies
acted upon by no net force, then what we call Newton's First Law
of Motion would be a genuine law with instances. This subjunctive
conditional claim, in turn, would be justified, since it would be
grounded on Newton's Second Law of Motion, which does have
instances. The upshot is that Armstrong's account of the nature of
laws can accommodate what might be referred to as *derived* laws
that lack instances. But what about *basic* laws? It is at this point
that the first difference between Armstrong's account and my own
emerges, since the latter allows for the logical possibility of basic
laws that lack (positive) instances, whereas the former does not.

A crucial issue, therefore, is whether it is possible for there to be
uninstantiated, *basic* laws. That this is possible is much less clear
than that there can be derived laws that are uninstantiated. I shall
argue, however, in section 3.1, that there is good reason to accept
the view that it is possible for there to be basic laws that have no
instances.

A second important difference concerns the number and variety
of contingent relations among universals that can be identified

with laws. According to Armstrong's theory, wherever one has a law, one has a *dyadic* relation between the corresponding universals. Moreover, the dyadic nomological relations form, so to speak, a single family: there is the relation of necessitation, and there are its probabilistic variants, for all probabilities greater than zero. This means that, probabilistic laws aside, all laws involve the single dyadic relation of necessitation.

When this theory of laws is combined with a theory of universals that rejects the possibility of negative and disjunctive universals, it follows that there are significant restrictions upon the logical form of underived laws. For example, it cannot be an underived law that having property P precludes the possession of property Q. Nor can it be an underived law that anything that possesses property P will also possess either property Q or property R.

These restrictions could be avoided by adopting a less austere theory of universals. If, however, as was suggested in Chapter 1, there is good reason for rejecting both negative universals and disjunctive universals, a different tack is called for if one is to have a theory that does not entail any restrictions upon the possible logical form of underived laws. In particular, as we shall see in section 2.2, one needs to hold both that laws may involve other dyadic relations between universals besides that of necessitation, and its probabilistic variants, and also that laws may involve relations among more than two universals—indeed, that there can be no a priori limit upon the number of universals that enter into a single, underived law.

The third fundamental difference concerns the character of basic laws. The difference emerges if one asks how a statement of a non-probabilistic law is related to the corresponding universal generalization. On the view to be advanced here, the relation is one of logical entailment: its being a law that Fs are Gs entails that all Fs, without exception, are Gs. The basic laws might, therefore, be referred to as *laws of strict necessitation*. Such laws entail that certain sorts of situations are nomologically impossible.

On Armstrong's account, by contrast, the basic laws are not laws of strict necessitation; they are rather what he refers to as *oaken laws*.[3] Such laws do not entail the existence of a corresponding, exceptionless regularity. Its being a law that Fs are Gs, on

---

[3] Armstrong, *What is a Law of Nature?*, pp. 147–50.

this view, rather than entailing that all *F*s are *G*s without exception, entails only that any *F* will be a *G*, *provided that* no interfering factor is present.

Given these fundamental notions, of oaken laws and of interfering factors that can block the operation of a law, one can go on to introduce concepts of more rigid sorts of laws, as Armstrong points out. If, for example, it is a law that *F*s are *G*s, and, in addition, there is as a matter of fact no property that is capable of blocking the action of that law, then one can speak of an *iron* law. Or if it is true not merely that there is no property that can interfere with the operation of the law, but that this is so by virtue of some second-order law, then one can speak of *steel* laws.[4]

It would seem, however, that there is a significant restriction with respect to the sorts of laws that can be introduced in this way. For although one can endow first-order laws with greater firmness by introducing higher-order laws, so long as all the laws in the resulting hierarchy are ones that necessitate a certain connection only in the absence of interfering factors, it would seem that however far the sequence of higher-order laws is extended, it will not suffice to make the laws on the bottom level equivalent, in effect, to laws of strict necessitation. For even if the hierarchy of laws were an infinite one, the fact that each law is, in itself, only one that necessitates if there is nothing that blocks its operation means that there could be a possible world that contained all of those laws, together with interfering factors for every one of them. In such a world it would be nomologically possible for something to be an *F*, while failing to be a *G*. The upshot seems to be that if one starts out from oaken laws, one cannot quite capture the idea of *nomological impossibility* which is associated with laws of strict necessitation.

By contrast, if one starts with the notion of laws of strict necessitation, one can introduce the idea of an oaken law, provided that one's account is sufficiently tolerant with respect to the logical form of basic nomological statements. Thus, for example, it will be an oaken law that *F*s are *G*s if there is some set of properties, *S*, such that it is a law of strict necessitation that if something is *F*, but has none of the properties belonging to *S*, then it also is *G*.

---

[4] Ibid., pp. 149–50.

What is the source of these differences? The answer is that they are all connected with the choice between, on the one hand, an Aristotelian theory of universals and, on the other, a theory that is at least partially Platonic in the sense of countenancing the possibility of uninstantiated universals. Armstrong's account of the nature of laws is a very systematic working out of what is, I believe, the most plausible view if one accepts a broadly Aristotelian theory of universals, whereas the present account represents an attempt to set out the sort of theory that naturally results if one is prepared to make some concessions in the direction of Platonism.

The main connections are roughly as follows: consider first the issue of the possibility of basic laws that never have instances. As we shall see shortly, one can certainly describe partial possible worlds of which it is at the very least tempting to say that such worlds contain basic laws that never have instances. There are, however, at least two reasons why an Aristotelian with respect to universals cannot be happy with this idea. In the first place, as will emerge in section 3.1, at least some of those partial possible worlds are such that an acceptance of underived laws without instances entails an acceptance of uninstantiated universals as well, given the assumption that laws are relations between universals. In the second place, even in the case of uninstantiated basic laws that do not involve uninstantiated universals, it seems likely that an Aristotelian cannot plausibly embrace the possibility of such laws, since it is hard to see what grounds could be given for countenancing a transcendent realm of basic laws without instances while rejecting a transcendent realm of uninstantiated universals.

This difference leads immediately to the next—that concerning the number and variety of nomological relations. Consider, for example, the question of whether there can be a triadic relation among the universals F, G, and H by virtue of which it is a law that if something has property F, but lacks property H, then it has property G. If laws and universals enjoy a mode of existence that transcends the existence of whatever instances they may happen to have, there would seem to be no problem concerning such a nomological relation. However the situation is rather different if one adopts the view that the existence of laws and universals is logically tied to their instances. For then one could have such a nomological relation only if it were possible for the law in question to exist, so to speak, *in* its instances. But instances of a law to the effect that if something has property *F*, but lacks property *H*, then it has property *G*, would have to be things that had properties *F* and *G*, but not property *H*, so it would appear to be impossible for any triadic relation among the universals *F*, *G*, and *H* to be present in the instances of such a law.

This in turn leads to the final difference—namely, that concerning whether oaken laws are basic. We have just seen that if one adopts an Aristotelian view one apparently cannot identify the law that *F*s which are not *H*s are *G*s with a corresponding triadic relation among the universals *F*, *G*, and *H*. What account, then, is to be offered of such laws? It would seem that there is no alternative to introducing a relation between those universals that are instantiated whenever one has an instance of the law—that is, between the universals *F* and *G*. Moreover, this dyadic nomological relation must be such that the fact that *F* and *G* stand in that relation does not entail that all *F*s are *G*s. Consequently, statements expressing oaken laws are not true by virtue of the existence of certain related laws of strict necessitation: their truthmakers are simple nomological states of affairs that have nothing to do with laws of strict necessitation. Oaken laws, therefore, are basic.

To sum up, the theory of the nature of laws to be proposed here is slightly more liberal than Armstrong's theory, in at least three respects. The fact that it is less restrictive with respect to the possible sorts of basic laws means, I believe, that it agrees more closely with our initial intuitions on this matter. However, as I have just indicated, that advantage has a price—namely, acceptance of a theory of universals that is at least partially Platonic. The question, therefore, is whether that price is unacceptably high.

Before we proceed, there are two terminological points that need to be mentioned. First, there is the question of how the term 'law' is to be employed. Some philosophers use the term 'law' to refer to certain sentences. Others use it to refer to certain propositions. And others use it to refer to objective states of affairs by virtue of which the corresponding sentences, or propositions, are true. I shall follow the latter usage. When I wish to refer instead to a sentence stating some law, I shall use expressions such as 'nomological statement' or 'law-statement'.

The second terminological point concerns the use of the expression, 'nomological statement'. Some writers use this expression to refer to statements such as 'For all x and y, if x exerts force F on y at time t, then the force which y exerts on x at time t is equal in magnitude to F and opposite in direction.' I shall instead refer to such a statement as a nomological *generalization*, or as a generalization that *expresses* a law, and I shall reserve the expression 'nomological statement' for statements that *explicitly* contain nomological terms, such as the statement 'It is a law that for all x and y, if x exerts force F on y at time t, then the force which y exerts on x at time t is equal in magnitude to F and opposite in direction.'

2.1 SOME UNSATISFACTORY ACCOUNTS OF LAWS

I shall begin be considering some important alternative accounts of the nature of laws. I think that a clear understanding of precisely how these accounts are defective will both point to certain conditions that any adequate account must satisfy, and provide strong support for the thesis that the truth-makers for nomological statements must be relations among universals.

2.1.1 *The Regularity Theory of Laws*

Most contemporary accounts of laws identify them with regularities. It is clear, however, that not just any regularity is a law. If Smith plants a garden, and is careful not to allow any fruit other than apples in that garden, a certain regularity will obtain—that described by the generalization, 'All the fruit in Smith's garden are apples'. This regularity would not be a law, since laws must support corresponding counterfactuals, and the regularity in question would not provide any ground for believing that if Smith had attempted to put an orange in the garden, he would have failed.

The problem, accordingly, is to find some way of distinguishing between regularities that are laws and those that are not. Contemporary defenders of a regularity approach have advanced various proposals. In what follows I shall examine the more important suggestions.

2.1.1.1 *Lawlike Generalizations*

Perhaps the most common version of the regularity approach to laws is the view that a generalization expresses a law if and only if it is both lawlike and true, where lawlikeness is a property that a statement has, or lacks, simply by virtue of its meaning. Different accounts of lawlikeness have been advanced, but one requirement is invariably taken to be essential: a lawlike statement cannot contain any reference, either explicit or implicit, to specific individuals. This requirement enables one to deal with generalizations such as 'All the fruit in Smith's garden are apples.' Since this statement entails the existence of a particular object—Smith's garden—it lacks the property of lawlikeness. So unless it is entailed by other true statements that are lawlike, it will be at best an accidentally true generalization.

Its success in handling this sort of case notwithstanding, this general approach is exposed to at least four serious objections. First, consider the statement 'All the fruit in any garden with property *P* are apples.' This generalization is free of all reference to particular individuals, so unless it is unsatisfactory in some other way, it will be lawlike. But *P* may be quite a complex property, so chosen that there is, as it happens, only one garden that ever possesses that property, namely Smith's. If that were so, one would not want to hold that it was a law that all the fruit in any garden with property *P* are apples. It seems clear, then, that generalizations can be both lawlike and true, yet fail to express laws.

A second objection to this approach arises in connection with what are usually referred to as 'vacuously true generalizations'. This label is not, however, an entirely happy one, since it suggests that the generalizations in question have no instances, that they are not true of anything at all. That this view of the matter is problematic becomes clear if one considers a typical example:

G: No unicorns have wings.

This generalization is of the sort that would be labelled 'vacuously true'. But is it really correct to say that this generalization has no instances, or that it is true of nothing at all? The problem emerges if one considers the contrapositive:

H: Nothing with wings is a unicorn.

One hardly wants to characterize this generalization as vacuously true, given that all of the many individuals with wings provide one with instances. But can *G* be vacuously true, if *H* is not, given that they are logically equivalent? Is it possible for one of two logically equivalent generalizations to be vacuously true, and the other not? It would seem not. The expression 'vacuously true' is surely intended as a semantical predicate, not a syntactical one, and this would seem to imply that, given two logically equivalent generalizations, either both are vacuously true, or neither is.

Rather than speaking of a generalization as being vacuously true, one might introduce the idea of *positive* instances of a generalization. But if this notion is to be a semantical one, one will have to classify both unicorns that lack wings and winged animals that are not unicorns as positive instances of the generalization, 'No unicorns have wings.' This means that one cannot say that the generalization in question has no positive instances. One will have to say instead that it does not have positive instances of all possible types.

This approach is perhaps satisfactory. There is, however, some question about the appropriate construal of 'positive instance' in this context, especially when one considers generalizations of a more complex logical form. For this reason, I am inclined to think that it may be best to employ the following approach. First, in order to explain what is meant by an instance of a generalization, let us consider Hempel's concept of the development of a hypothesis with respect to a finite class of individuals. Expressed informally, a sentence *S* represents the development of a hypothesis *H* with respect to some class of individuals, *C*, if and only if *S* 'states what *H* would assert if there existed exclusively those objects which are elements of *C*'.[5] Thus, for example, if *H* is the hypothesis '$(x)(Fx \supset Gx)$', and *C* is the class containing only two individuals, *a* and *b*, then the development of *H* with respect to *C* will be given:

$$(Fa \supset Ga) \ \& \ (Fb \supset Gb).$$

Given this notion, one can then say that a state of affairs, *T*, involves an instance of a generalization, *H*, if and only if there is

[5] Hempel, 'Studies in the Logic of Confirmation', *Mind*, 54, 1945, pp. 1–26 and 97–121. Reprinted in Hempel, *Aspects of Scientific Explanation*, New York, 1965, pp. 3–51: see p. 36.

some individual, *a*, which is an element of *T*, and some sentence *S* which, first, is true by virtue of *T*, and second, which represents the development of *H* with respect to the class whose only member is *a*. Next, given this notion of an instance of a generalization, one can go on to talk about different types of instance of a given generalization. Thus, for example, in the case of simple conditional generalizations, such as $(x)(Fx \supset Gx)$, there will be three possible types of instance:

(1) Cases where something is both an *F* and a *G*.
(2) Cases where something is a *G*, but not an *F*.
(3) Cases where something is neither an *F* nor a *G*.

Finally, given these notions, one can then speak, not of a generalization's having, or failing to have, positive instances, but simply of a generalization's having, or failing to have, instances of all possible types. And when, in what follows, I speak of a generalization as being vacuously true, it is this property to which I am referring.

The generalization 'No unicorns have wings' can thus be classified as vacuously true, for it is a simple conditional generalization that has instances of types (2) and (3), but not of type (1).

Let us now turn to the difficulty: What is the status of a generalization such as 'No unicorns have wings'? It certainly appears to be lawlike, and it is true under the usual interpretation. But surely it does not express a law. For in the first place, the fact that there are no unicorns, and *a fortiori*, no unicorns with wings, provides one with no reason at all for holding that if there *were* unicorns, they *would* not have wings. The generalization in question does not support corresponding counterfactuals. Secondly, if the generalization, 'No unicorns have wings', were treated as expressing a law, the same would have to be done with the generalization, 'No unicorns lack wings.' It would then follow from these two nomological generalizations that the existence of unicorns is nomologically impossible—a conclusion which is very counterintuitive.

How is this difficulty to be avoided, if one adopts the view that generalizations express laws if they are both lawlike and true? There seem to be three main proposals. One involves the idea that the generalization in question cannot be vacuously true. A second approach appeals to some condition concerning the types of instance that a generalization has. A third suggestion is that a generalization cannot express a law if it contains some term—such as 'unicorn'—that fails to apply to anything. The third suggestion can be dealt with quite quickly. In the first place, there are generalizations that satisfy that requirement, but still seem to give rise to what appears to be the same basic problem. Consider, for example, the generalization:

> Whenever two lumps of gold, each weighing more than one million tons, come into contact, they turn red.

One does not want to say that this expresses a law, even though it is not only lawlike and true, but free of terms that fail to refer to anything. In the second place, the suggested requirement is not satisfactory even in the unicorn case, since one can replace the term 'unicorn' by some *analysans*, thereby generating an analytically equivalent generalization that is free of terms that do not apply to anything.

This leaves us with proposals one and two. I have already argued, however, that it will not do to appeal to the idea that a generalization cannot express a law if it is vacuously true. We need only consider, therefore, the idea of appealing to some restriction formulated in terms of the types of instance that a generalization has.

The most plausible version of such a condition is, I think, as follows: If a generalization has instances of all possible types, then it expresses a law if it is both lawlike and true. But if it does not have instances of all possible types, more is required. It expresses a law only if it is derivable from other generalizations that are lawlike and true, and that do have instances of all possible types.

I believe that it can be shown, however, that this requirement is not acceptable. Imagine a world containing ten different types of fundamental particle, and let us suppose that the behaviour of particles when they collide depends upon the types of the particles involved. If one considers only interactions involving two particles, there are fifty-five possibilities with respect to the types of the two particles. Suppose that collisions involving particles of types X and Y have not been studied, but that the other fifty-four possibilities have been subjected to close experimental scrutiny, with the result that fifty-four laws have been discovered, one for each case. Let us suppose, finally, that each of these laws is basic, so that it is not possible to derive even two of them from some simple, more general law. In this situation, what would it be reasonable to believe regarding collisions involving particles of types X and Y? I suggest that it would be reasonable to believe that such interactions were also governed by some law, and one, moreover, which was also basic.

What if it turns out, however, that particles of types X and Y have never collided at any time? Would that change the situation? In particular, would it then no longer be reasonable to believe that there was a fifty-fifth underived law governing the interaction of particles of types X and Y? It is hard to see why this would be the case.

This brings us to the final step. Suppose that in addition to the fact that particles of types X and Y have never collided at any time, it is also the case that the world, though not completely deterministic, is sufficiently so that when a grand Laplacean calculation is performed, it turns out that, given the current distribution of particles, it will be impossible for particles of types X and Y to interact at any time in the future. Would it still be reasonable to hold that there was probably some underived law governing the interactions of particles of types X and Y?

One way of thinking about the question is this: Since the world we are imagining is not completely deterministic, we can suppose that there was a certain particle of type Z that could have veered to the left rather than to the right, and that if it had veered to the left, that would have led to innumerable interactions involving particles of types X and Y. Now given the existence of fifty-four underived laws governing the other types of interactions, it seems very reasonable to hold that if X and Y particles had interacted, their interactions would have been governed by another underived law. But does one want to say, then, that the fact that a certain particle veered to the right, rather than to the left, made it the case not merely that there were no instances of a law governing interactions of particles of types X and Y, but no law at all? I believe that there is quite a strong intuition that what laws there are in a world should not depend upon chance events in this way. If this intuition is justified, then there would seem to be good reason for holding that there would be an *underived* law governing the interaction of particles of types X and Y even in a world in which, by misfortune, they happen never to meet.

But it is precisely this view that has to be rejected if one accepts the claim that a generalization that fails to have instances of all possible types can express a law only if it is derivable from generalizations that do satisfy this requirement. We have, then, a reason for rejecting that claim.

The thesis that it is logically possible for there to be basic laws that fail to have instances of all possible types is a crucial one, for at least three reasons. First, the possibility of vacuously true, *underived* nomological statements appears to be incompatible with *any* regularity theory of laws. Second, this possibility seems to provide very strong support for the theory that laws are relations among universals, as I shall argue shortly. Finally, as mentioned before, it bears upon the choice between Armstrong's version of that theory and the one being proposed here.

Given that this is so, let me try to offer some additional support for the claim that such laws are in fact possible. One argument, which does involve a controversial assumption, may be put as follows. There are good reasons for believing that there is some very intimate relation between certain neurophysiological states on the one hand, and experiences on the other. Some philosophers hold that the relation is actually that of identity. For the sake of the present argument, however, let us assume that reductive materialism is false. Granted this assumption, it is very plausible that the relationship in question involves the existence of psychophysical laws linking neurophysiological states, as causes, to corresponding experiences, as effects. Given certain neural states, a person will have an experience of the red variety. Given other neural states, an experience of the green variety, and so on.

Next we must ask about the status of such laws. Are they basic, or derived? It is possible that some of the psychophysical laws are derived, but if they are, it can only be from other psychophysical laws, given the assumption that reductive materialism is not the case. Therefore at least some psychophysical laws, if not all, must be basic.

The next step in the argument involves considering what the world would have been like if our earth had been slightly closer to the sun, and if conditions in other parts of the universe had been such that life evolved nowhere else. The universe would then have contained no sentient organisms, and *a fortiori*, no experiences of the red variety. But would it not have been true *in that world* that if the earth had been a bit further from the sun, life might very well have evolved, in which case there would have been experiences of the red variety? If so, by virtue of what would this counterfactual have been true? Surely an essential part of what would have made it true would have been the existence of a psychophysical law linking complex physical states to experiences of the red variety. But the world we are imagining contains no experiences of the red variety. Therefore the psychophysical law in question could not have instances of all possible types. Moreover, in the world we are considering, the same thing will be true of absolutely all psychophysical laws. It follows, therefore, that there would be, in that world, *basic* laws that are expressed by vacuously true statements.

What the argument shows, in short, is this. If reductive materialism is false, there is a possible world that differs from the actual world only with regard to the 'initial distribution' of matter and energy, and in which there are basic laws that are expressed by vacuously true generalizations.

What if one rejects the assumption that reductive materialism is false? The argument will then have to be reformulated, with the assumption that reductive materialism is false replaced by the assumption that it is at least logically possible that reductive materialism is false. With this modification, it will no longer be possible to argue that there is a possible world that differs from the actual only with respect to the initial distribution of matter and energy, and in which there will be underived psychophysical laws that fail to have instances of all possible types. However the revised argument will still lead to the conclusion that it is possible for there to be worlds containing psychophysical laws that are expressed by vacuously true generalizations.

To sum up, I believe that the two cases set out above—the case of the particles that happen never to meet, and the case of psychophysical laws in a world that happens not to contain any sentient beings—lend strong support to the thesis that it is logically possible for there to be *underived* laws that fail to have instances of all possible types. Other things being equal, then, an approach that allows for the possibility of basic laws that are expressed by vacuously true generalizations is to be preferred to an approach that does not. It could turn out, of course, that other things are not equal. Perhaps any theory of laws that permits there

The Basic Approach 51

to be underived laws that fail to have instances of all possible types necessarily involves some other defect, in which case the intuition that there can be such laws might have to be abandoned as ultimately mistaken. I shall attempt to show, however, that it is possible to formulate a theory that does permit the existence of such laws without being exposed to other, even more serious, difficulties.

A third objection to the view that lawlike generalizations express laws is due to William Kneale. In his articles, 'Natural Laws and Contrary-to-Fact Conditionals', and 'Universality and Necessity',[6] Kneale points out that we naturally believe that there may be any number of types of situation that are nomologically possible, but which are never in fact realized:

. . . let us suppose that a musician composes an intricate tune in his imagination while he is lying on his death bed too feeble to speak or write, and that he says to himself in his last moments 'No human being has ever heard or will ever hear this tune,' meaning by 'this tune' a certain complex pattern of sounds which could be described in general terms. Obviously he does not think of his remarks as a suggestion of natural law.[7]

But if lawlike generalizations express laws, it is logically impossible for there to be any unrealized possibilities of this sort. For if the unrealized situation is of a type that can be described in purely general terms, the fact that such a situation never occurs means that a corresponding lawlike generalization is true, so if lawlike generalizations express laws, it follows that the unrealized situation in question is in fact nomologically impossible. This conclusion— that it is logically impossible for there to be any general, unrealized possibilities—seems extremely counter-intuitive.

A fourth objection to the view that laws can be explained in terms of lawlike generalizations is this. Assuming that there can be statistical laws, let us suppose that it is a law that the probability that something with property P has property Q is 0.999999999. Suppose further that there are, as a matter of fact, very few things in the world with property P, and, as would then be expected, it

---

[6] Kneale, 'Natural Laws and Contrary-to-Fact Conditionals', *Analysis*, 10/6, 1950, pp. 121–5, and 'Universality and Necessity', *British Journal for the Philosophy of Science*, 12/46, 1961, pp. 89–102. Both are reprinted in *Philosophical Problems in Causation*, ed. Beauchamp, Encino and Belmont, Calif., 1974.

[7] Kneale, 'Natural Laws and Contrary-to-Fact Conditionals', in Beachamp (ed.), ibid., p. 49.

happens that all of these things have property Q. The statement that everything with property P has property Q would then be both lawlike and true, yet it would not express a law.

One might even have excellent grounds for holding that it was not a law. For there might be some powerful and well-established theory which entailed that the probability that something with property P would have property Q was not 1.0 but 0.999999999, thus implying that it was not a law that everything with property P would have property Q.

If this final argument is correct, it shows something rather important: namely, that there are statements that would express laws in some worlds, but be merely accidentally true generalizations in others. So there cannot be any property of lawlikeness which a statement has simply by virtue of its meaning, and which together with truth is sufficient to make the statement one that expresses a law. The 'lawlike-generalizations' approach to laws must therefore be wrong in principle, and not merely with respect to certain details.

The above approach to laws is also exposed to a number of other important objections.[8] I believe, however, that the above objections are very strong, and that they suffice to show that there is little hope of defending the present formulation of the regularity approach to laws.

2.1.1.2 *Structural or Systematic Approaches*
One of the most important approaches to the problem of formulating a more satisfactory version of the regularity theory turns upon the idea that one needs to consider complete systems of generalizations if one is to be able to distinguish between laws on the one hand, and accidentally true generalizations on the other. I shall consider two variations on this general theme—the one due to Karl Popper, and the other to F. P. Ramsey and David Lewis.

Popper's Non-Deducibility Requirement. Popper's approach is set out in his essay, 'A Revised Definition of Natural Necessity',[9] and may be characterized informally as follows: on the one hand,

---

[8] For a very thorough discussion of objections to this approach, see chaps. 2–4 of Armstrong, *What Is a Law of Nature?*

[9] Popper, 'A Revised Definition of Natural Necessity', *British Journal for the Philosophy of Science*, 18, 1967, pp. 316–21, and reprinted in Beauchamp (ed.), op. cit., pp. 66–72.

Popper holds that a generalization cannot express a law unless it is both lawlike and true: on the other, he recognizes that these conditions are not sufficient. The question, then, is whether there is any further requirement that can be imposed which, together with lawlikeness and truth, will suffice to ensure that a generalization expresses a law.

Popper believes that there is, and he sets out a formal account of this additional requirement. His proposal can probably best be grasped, however, by considering an example such as his own case of the moas.[10] These were large birds which once lived in New Zealand but are now extinct. Popper asks us to make two suppositions. First, that no moas ever have existed, or ever will exist, other than those that lived in New Zealand. Second, that although the biological make-up of moas was such as to permit them to live well beyond the age of fifty, as a matter of fact none of the moas in New Zealand did live beyond the age of fifty, due to a certain virus which happened to be present at that time. Given these suppositions, it would be true that no moa lives beyond the age of fifty. Yet one would not regard this as a law of nature. Perhaps, then, if we can clarify why we would not regard this as a law, we will have isolated the additional condition that must be added to lawlikeness and truth if a generalization is to express a law.

Why would we not regard it as a law, in the situation described, that all moas die before the age of fifty? The natural response centres upon the point that this generalization concerning moas is *derivable* from certain other generalizations taken in conjunction with what might be called initial and/or boundary conditions. Given that there were certain sorts of viruses in New Zealand at the time of the moas, and given that moas have not, and will not, exist at any other time or place, then, by virtue of appropriate generalizations concerning the effects of a certain virus upon animals with a specific biological make-up, it follows that there will not be any moas that live beyond the age of fifty. And it is this derivability, surely, that leads one to classify the generalization concerning moas as one that is merely accidentally true.

Let us say, then, that a generalization satisfies the non-deducibility requirement if it is not derivable from true statements,

10 Popper, *The Logic of Scientific Discovery*, London, 1961, pp. 427–8.

some of which are statements of initial and/or boundary conditions, and others of which are lawlike generalizations. The question, then, is whether the following thesis is true:

> A generalization expresses a law if and only if it is lawlike and true, and satisfies the non-deducibility requirement.

I shall argue that there are at least three reasons for rejecting this claim. The first objection arises because of the logical possibility that reality is not completely deterministic. Consider a generalization such as the following: 'Every electron is within 100 light-years of some proton.' This generalization could be true in our own world, but if it were, we would not be likely to classify it as one that expresses a law. Let us suppose, then, that this generalization is true, but not nomological. Will it be possible to deduce it from some statements describing initial and/or boundary conditions together with nomological statements? It seems to me that this need not be the case. If the world in question is completely deterministic, then any set of statements containing all nomological statements, together with statements describing completely what the world is like at any one moment, will entail all truths about the world, and so will settle what non-nomological generalizations are true in that world. But on the other hand, if the world is not completely deterministic, then there need be no set of statements describing initial and/or boundary conditions that, when taken together with the set of nomological generalizations, will entail all true non-nomological generalizations. Perhaps, in the case we are considering, it is true that all electrons are within 100 light-years of some proton, but that if one particular proton had veered to the left at a certain time, this would not have been the case. In an indeterministic world, then, there could be generalizations that were lawlike and true, and which satisfied the non-deducibility requirement, but which did not express laws.

The second objection is this: imagine a world in which, first, it is a law that if two particles of type P collide when the velocity of each is less than 0.999 of the velocity of light, an event of type R will occur; second, it is a law that if two particles of type Q collide, whatever their velocities, an event of type R will occur; and third, that it is not a law that particles of type Q need be particles of type P. Finally, let us suppose that the following two things must be the case, given the *other* laws that there are in the world, taken together with a complete description of the world at some moment: (1) all particles of type Q will also be particles of type P, and (2) no particles of type Q will ever travel faster than 0.999 of the velocity of light. Given these stipulations, the generalization 'Whenever particles of type Q collide, an event of type R occurs' will be derivable from other generalizations, given statements describing appropriate initial conditions. If the non-deducibility requirement were correct, the generalization in question could not possibly express a law in the world described—a conclusion which seems counter-intuitive.

The third objection concerns the case of underived, vacuous laws. The requirement of non-deducibility provides no assistance in handling that case. For consider again the universe where particles of types X and Y happen never to meet, even though they could. If the generalization 'Whenever particles of types X and Y collide an event of type E occurs' fails to satisfy the non-deducibility requirement in the world in question, then it will not be a law—a conclusion which, it was argued above, does not seem plausible. On the other hand, if it is the case—as seems more likely—that the generalization does satisfy the non-deducibility requirement, then the same will also be true of a generalization such as 'Whenever particles of types X and Y collide, an event of type F occurs, rather than one of type E.' Since both generalizations are also lawlike and true, both must be laws. From this it follows that it is, in the world in question, a law that particles of types X and Y can never collide—a consequence which seems unacceptable.

The Ramsey/Lewis Approach. Another structural approach to the problem of analysing the concept of a law of nature along regularity lines is due to Ramsey, who suggested that the generalizations that express laws are those that are 'consequences of those propositions which we should take as axioms if we knew everything and organized it as simply as possible in a deductive system'.[11]

This general view is embraced by David Lewis, at least as a working hypothesis, in his book, *Counterfactuals*. The formulation that he offers is: 'A contingent generalization is a *law of nature* if

[11] Ramsey, 'General Propositions and Causality', *The Foundations of Mathematics*, ed. Braithwaite, 1960, pp. 237–55. See p. 242.

and only if it appears as a theorem (or axiom) in each of the true deductive systems that achieves a best combination of simplicity and strength.'[12] This formulation is an improvement upon Ramsey's, both because of elimination of the unnecessary counterfactual about omniscience, and because of Lewis's stress upon the possible conflict between the virtue of simplicity and that of strength.

I believe that this approach is one of the two most satisfactory versions of the regularity theory. None the less, it is still exposed to some very serious objections. In the first place, there is once again the problem of underived, vacuous laws. Suppose it is a law in our world that whenever an organism goes into a brain state of type B, it has a visual experience of the purple variety, and consider what would have been the case if all life had been destroyed before any sentient being had seen a purple object. On the Ramsey/Lewis approach, the generalization 'Whenever an organism is in a brain state of type B, it has an experience of the purple variety' would not express a law in that world, since the addition of that generalization to a deductive system would decrease the simplicity of the system without increasing its strength. Similarly, in the world where particles of types X and Y happen not to meet, there would not be any law governing how they would have behaved if they had happened to collide.

Secondly, one of the objections to Popper's approach also tells against the Ramsey/Lewis approach, namely, the case where the truth of a generalization that expresses a law happens to be derivable from statements of other laws together with statements of initial conditions. When this situation obtains, inclusion of the generalization in the overall theory would increase the complexity of the theory without increasing its strength. The present approach would therefore imply that the generalization in question was not nomological.

A third objection is this. Imagine a world in which some generalization, G, gets classified as a law because it belongs to each of the true deductive systems that best combine strength and simplicity. If the initial conditions in the world had been slightly different, the law that is expressed by G might have had more instances, or it might have had fewer. Consider, then, a sequence

---

[12] Lewis, *Counterfactuals*, Cambridge, Mass., 1973, p. 73.

of possible changes in initial conditions that would generate a sequence of possible worlds in which there would be fewer and fewer instances of the law expressed by generalization *G*. On the Ramsey/Lewis approach it would seem that a point will be reached at which *G* will no longer express a law, since the increase in strength due to its presence in the deductive system will not be sufficiently great to outweigh the loss in simplicity. And this consequence, I suggest, conflicts with a rather firm intuition concerning the nature of laws, to the effect that what laws there are should not be dependent upon such accidental factors as the number of instances the relevant generalization happens to have.[13]

2.1.1.3 *Laws and Supporting Evidence*
Whether we regard a generalization as nomic, or as merely accidental, makes a great deal of difference with respect to the ways in which we are willing to employ the generalization. If we view a generalization as expressing a law, we advance counterfactuals that are based upon it, we employ it in offering explanations, and we use it in making predictions, all of these to an extent and in ways that we would not dream of doing if we regarded the generalization as merely accidental. Given that this is so, might it not be possible to develop an account of the difference between laws and accidentally true generalizations on the basis of these differences in epistemic attitudes?

On the face of its, this approach would not seem very promising. For in the first place, if one says that what makes it the case that a generalization expresses a law is that people have certain epistemic attitudes towards it, what is one to say about the possibility of unknown laws? Or about the logical possibility of a universe that is governed by laws, but which never contains any intelligent life? In order to accommodate such possibilities it seems that one will have to refer not merely to actual epistemic attitudes, but also to the attitudes that people *would* have if certain things were the case. This, however, introduces a circularity into the analysis, since it seems very unlikely that a satisfactory account of the truth

[13] With regard to the last point, compare Suchting's discussion on pp. 81–2 of 'Regularity and Law', *Boston Studies in the Philosophy of Science*, 14, ed. Cohen and Wartofsky, New York, 1974, pp. 73–90. For a more thorough discussion of objections to the Ramsey/Lewis systematic approach, see Armstrong, *What Is a Law of Nature?*, chap. 5.

conditions of counterfactuals can be offered which does not involve reference to laws.

Secondly, unless one is prepared to abandon a realist view of laws, it will not do to explain what it is that makes a generalization nomological on the basis of the epistemic attitudes that people happen to have. If the distinction between nomic generalizations and merely accidental ones is to correspond to something in the world external to us, then one must shift from talk about the epistemic attitudes that people actually have to talk about the epistemic attitudes that it is rational to have, given the way the world is.

This second observation points towards a way in which the above approach can be transformed, and which enables one to avoid the problems of unknown laws, and mindless universes. Rather than talking about epistemic attitudes, talk instead about the facts that provide a *ground* for the relevant attitudes—the considerations that make certain attitudes reasonable ones. Whether a generalization expresses a law, then, will be a matter of whether the generalization is supported by a certain sort of evidence—evidence that would make it reasonable to treat the generalization in certain ways. Such evidence might exist even if people were not aware of it, or failed to notice that it made it reasonable to treat a certain generalization as nomic, or even if the universe were one that never contained any intelligent beings.

The idea that the difference between nomological generalizations and accidental ones may be a matter of a difference in the supporting evidence is a familiar one. Many writers have pointed out, for example, that in the case of generalizations that express laws, in contrast to accidentally true generalizations, our supporting evidence typically involves a wide variety of instances. But perhaps the most appealing formulation of this general view is one that makes use of what Brian Skyrms has called the concept of *resiliency*.[14]

The notion of resiliency may be construed either epistemically or objectively. For our purposes, the relevant concept is that of epistemic resiliency, which can be explained as follows: suppose that a proposition, q, has probability m relative to some body of

[14] The concept of resiliency is set out by Skyrms in 'Resiliency, Propensities, and Causal Necessity', *Journal of Philosophy*, 74, 1977, pp. 704–13, and more recently in greater detail in *Causal Necessity*, New Haven and London, 1980.

evidence. The probability that q is the case may very well vary when the original evidence is supplemented in various ways, and the concept of epistemic resiliency is intended to provide a measure of the stability of the original probability. How should this be done? One possibility would be to try to define what might be called an absolute notion of epistemic resiliency by considering how the probability of q can vary with respect to every possible way of supplementing the evidence. Skyrms feels, however, that this approach is problematic, so instead he defines a relative notion of epistemic resiliency, which incorporates a specification of possible additions to the evidence to be considered. In particular, Skyrms suggests that one sets out some list of factors— $p_1 \ldots p_n$—and then considers how the probability of q differs from m when the original evidence is supplemented by different truth-functional combinations of the factors $p_1 \ldots p_n$—ignoring, of course, any truth-functional combinations that are logically incompatible either with q or its negation, since such combinations will force the probability of q to be either zero, or one, respectively. If the probability of q does not vary at all when the original evidence is supplemented by any truth-functional combination of the factors $p_1 \ldots p_n$ compatible both with q and its negation, then the proposition that the probability of q is equal to m exhibits perfect epistemic resilience with respect to the factors $p_1 \ldots p_n$, and will be assigned a resilience of one. If, on the other hand, the probability does vary, then the epistemic resiliency is to be determined, Skyrms suggests, by the maximum divergence from m under some supplementation of the original evidence by admissible truth-functional combinations of factors $p_1 \ldots p_n$. In particular, Skyrms suggests that one might take (1 − the maximum deviation from m) as an appropriate measure of the resiliency of the proposition that the probability of q is equal to m.[15]

How is this notion of epistemic resiliency to be applied to laws? The answer emerges if one focuses upon a point that is stressed by Skyrms in his discussion of non-statistical laws in chapter two of *Causal Necessity*, namely, that any satisfactory account of laws should enable one to make sense of the fact that laws can be used to make predictions: if one has reason to believe that it is a law that $(x)(Fx \supset Gx)$ and also that Fa, then one has reason to believe

---

[15] Compare Skyrms's definition of resiliency, in *Causal Necessity*, pp. 11–12.

that *Ga*. What condition might be imposed upon a generalization
to ensure that this is the case? A possible answer is that the
generalization, in order to be classifiable as a law, must be very
probable relative to one's evidence. But this suggestion is exposed
to the difficulty that if the universe is known to be very large, then
any contingent generalization will get assigned a low probability,
at least by standard systems of inductive logic.[16]

The natural response to this difficulty is to shift from considering
the probability of a generalization to considering the probability of
its instances. For in deriving the conclusion that *Ga* from the
premise that *Fa* one does not have to employ the generalization
that $(x)(Fx \supset Gx)$. One can appeal instead to the relevant material
conditional, $Fa \supset Ga$, which may have a high probability even if
the generalization $(x)(Fx \supset Gx)$ does not.

The defect in this suggestion emerges if one considers the
generalization that all unicorns have wings. The associated
material conditional 'If *a* is a unicorn, *a* has wings' is very probable
relative to the evidence that one now has, since that evidence
makes it very unlikely that any particular thing, *a*, chosen at
random, will be a unicorn, and so makes it very likely that *a* will
either not be a unicorn or will have wings. But if one were to learn
that *a* is in fact a unicorn, it would not be reasonable to draw the
conclusion that *a* has wings. So high probability of the relevant
material conditional does not suffice to ensure that one can employ
the generalization in making predictions.

At the same time, this example points in the direction of a more
promising account. For the problem appears to be that while the
probability of the material conditional 'If *a* is a unicorn, then *a* has
wings' is high relative to our evidence, it will not remain high if
that evidence is supplemented in certain ways—and in particular,
if it is supplemented by the proposition that *a* is a unicorn. Perhaps
what is needed, then, is not merely that the probability of the
material conditional be high, but that this probability be a resilient
one.

This suggests the following account: if a generalization is to
express a law, then it must be the case that there are propositions
describing particular states of affairs relative to which any

---

[16] Some philosophers have thought that this shows that such systems of inductive
logic are defective. I shall argue later, however, that this view of the matter is
mistaken.

instantiation of the generalization has a probability of one. This, however, is clearly not sufficient, since this condition will be met by any true generalization. One must require, in addition, that that probability of any instantiation of the generalization remains at least close to one when the original propositions are supplemented by other propositions, including ones that may be false in the world as it is.[17]

This idea—of explaining the difference between generalizations that express laws, and those that are merely accidentally true, in terms of the resiliency of the probability associated with instantiations of the generalization—is an interesting one. It does enable one to deal with certain objections to simpler versions of the regularity theory. Recall, for example, the case where there is a garden that contains only apples, and one chooses some highly specific property *P* that is possessed by that garden, but by no other, at any time. It will then be true, but not a law, that all the fruit in any garden with property *P* are apples. This case poses no difficulty for the present approach. For while it is true, for any *a*, that the probability of the material conditional 'If *a* is a garden with property *P*, then *a* contains no fruit other than apples' is high, this probability is not resilient with respect to relevant ways of supplementing the evidence. Given the assumption for example, that there are other gardens with property *P*, the material conditional in question would no longer have a high probability.

There are, however, decisive objections, not only to the attempt to explain the concept of a law in terms of the concept of resiliency, but to the general idea that the distinction between generalizations that express laws and those that are merely accidentally true can be explained in terms of the sorts of actual states of affairs which support the generalization. In the first place, consider the following variation on an earlier example. It seems possible that one might be in a world where it was a law that the probability that something which had property *P* had property *Q* was equal to 0.9, but in which it happened that everything with property *P* also had property *Q*. Moreover, this could occur even if there were a large number of things, of different sorts, which

---

[17] Restrictions will have to be placed, of course, upon what supplementary propositions are admissible. Skyrms advances what seems to be a plausible suggestion on p. 36 of *Causal Necessity*. The details are not crucial, however, in the context of the present discussion.

possessed property P. But in that case, not only would the material conditional 'If a has property P, then a has property Q' receive a high probability, that probability would be resilient relative to appropriate additions to the evidence. The generalization that all things that have property P have property Q would therefore have to be regarded, on the present approach, as expressing a law.

Secondly, there is once again the problem of underived, vacuous laws. If the distinction between nomological generalizations and accidental ones is a matter of the type of supporting evidence, then there is no possibility of there being a law dealing with the interaction of types of particles that happen never to meet, nor of there being psychophysical laws in the universe where sentient life happens not to evolve. The corresponding generalizations will have to be treated as on a par with the generalization that all unicorns have wings. If my earlier argument was right, this is strongly counter-intuitive.

Thirdly, the present approach is also exposed to the final objection directed against the systematic approach in the previous section. Suppose, for example, that one is in a world where not only is it true that all the fruit in any garden with property P are apples, but where the generalization has a large number and wide variety of instances, and has been repeatedly put to the test. People have tried to carry oranges into such gardens, and to grow pears in them, and have always failed. The material conditional 'If a is a piece of fruit in any garden with property P, then a is an apple' will then be very probable, and that probability will have high resiliency. So on the present account, that generalization must express a law. But now imagine the world being gradually altered, so that due to changed initial conditions, there are fewer and fewer instances of the generalization, and fewer and fewer situations in which the generalization is put to the test. Given the present approach, one would be forced to say that a point will be reached at which the generalization will not express a law. Thus a generalization may be disqualified from being nomological by accidental features of initial conditions which, though perfectly compatible with the generalization, have the result that the generalization has very few instances. This dependence of laws upon initial conditions, and upon the number and/or distribution of instances seems very counter-intuitive.

It would seem, then, that the theory that laws are to be

identified with regularities cannot be saved by appealing to some condition concerning the type of supporting evidence. This conclusion, together with the earlier ones to the effect that the distinction between accidental and nomological generalizations cannot be explicated either in terms of some property of lawlikeness, or in terms of structural interrelations among generalizations, strongly suggests that alternatives to the regularity theory need to be considered. For it is hard to see any other promising ways in which one might attempt to make that theory viable.

2.1.2 *Laws and Generalized Subjunctive Conditionals*

Law-statements entail corresponding counterfactuals, and support others, in ways in which accidentally true generalizations do not. This fact is often cited as an objection to the regularity theory of laws. For why should the fact that all Fs have happened to be Gs provide one with any reason for concluding that some a, which is neither an F nor a G, would be a G if it were an F?[18]

Though this objection is initially plausible, especially against the simple regularity theory, I am not convinced that it tells against more sophisticated versions. John Mackie has argued, for example, that if one first adopts the view that laws are expressed by generalizations that have good inductive support, and second, treats counterfactuals not as statements, but as condensed arguments, then one can give a satisfactory account of the relations between laws and counterfactuals.[19] The view of counterfactuals which Mackie proposes is not the one which I should be inclined to adopt. But on the other hand, I am not convinced that there is any decisive objection in it. If that is so the above objection to the regularity theory of laws may have to be set aside.

The question that I wish to consider in this section, however, is whether one can explain the relationship between laws and counterfactuals in a different way—namely, by offering an account of laws in terms of counterfactuals.

On the face of it, such an approach would not seem especially promising. For on the one hand, given that the truth conditions of

---

[18] Compare Kneale's discussion in *Probability and Induction* Oxford, 1949, sect. 17, pp. 74–5.

[19] Mackie, 'Counterfactuals and Causal Laws', *Analytical Philosophy*[1], ed. Butler, Oxford, 1966, pp. 65–80.

counterfactuals seem to be even less clear than those of law-statements, it will hardly do to treat counterfactuals as primitive statements not needing any analysis. And on the other hand, it is not easy to see how counterfactuals can plausibly be analysed without making use of the notion of a law. For consider the two main approaches to counterfactuals—the traditional, consequence approach, and the more recent, possible worlds alternative. The former approach makes explicit use of the notion of a law, since it says, very roughly, that the counterfactual 'If *p* were the case, then *q* would be the case' is true if and only if (1) *p* is false, and (2) *p*, when conjoined with some true law-statements, possible together with some other true statements which meet certain conditions, logically entails *q*. The possible worlds approach, in contrast, in claiming that truth conditions for counterfactuals can be given in terms of similarity relations over possible worlds, does not immediately bring in any reference to laws. It would seem, however, that the concept of a law will have to be brought in at a later point in the analysis, since one needs to clarify the factors that enter into judgements concerning the relative similarity of different possible worlds, and it seems very plausible—as Lewis argues in his own exposition of the possible worlds approach—that one of the factors that will weigh most heavily with respect to the determination of the degree of similarity of different possible words is the extent to which those worlds possess the same laws.[20]

None the less, the idea that law-statements can be analysed in terms of subjunctive conditionals has been proposed. In their article, 'A Semantic Analysis of Conditional Logic', Robert Stalnaker and Richmond Thomason offer a formal account of a possible worlds approach to subjunctive conditionals. They then suggest that a statement such as that it is a law that all *P*s are *Q*s can be analysed as equivalent to the generalized subjunctive conditional that for all *x*, if *x* were a *P*, then *x* would be a *Q*.[21]

This approach to laws is exposed to at least three objections. First, there is the point mentioned above, namely, that it would seem that any satisfactory measure of the similarity of different possible worlds will have to take into account the extent to which the worlds share the same laws, which implies that the analysis of

---

[20] Lewis, op. cit., pp. 74–5.
[21] Stalnaker and Thomason, 'A Semantic Analysis of Conditional Logic', *Theoria*, 36, 1970, pp. 23–42. See pp. 39–40.

laws proposed by Stalnaker and Thomason is implicitly circular. Second, there is the question of whether the possible worlds approach to subjunctive conditionals is itself satisfactory, or whether it does not in fact have to be abandoned in favour of the traditional, consequence analysis, which makes explicit reference to laws. One problem for a possible worlds approach, for example, is this: consider a world in which neither events of type *Q* nor events of type *R* are causally determined, but where it is a law that every event of type *P* is followed either by an event of type *Q* or by an event of type *R*. Let *S* be some appropriate spatio-temporal region that contains neither events of type *P*, nor events of type *Q*, nor events of type *R*. Assume, finally, that while events of type *Q* are minor occurrences indeed, events of type *R* have cataclysmic consequences. Given these assumptions, what truth value is to be assigned to the following counterfactual: 'If an event of type *P* had occurred in region *S*, it would have been followed by an event of type *Q*, not by an event of type *R*'?

On the traditional, consequence approach, this counterfactual would be false. What would be true would be the weaker counterfactual to the effect that if an event of type *P* had occurred in region *S*, then that would have been followed either by an event of type *Q* or by one of type *R*. But on the possible worlds approach, the stronger counterfactual would be true. For one has to consider, according to the Stalnaker/Thomason approach, that possible world which, subject to the constraint that it is a world in which region *S* contains an event of type *P*, is most similar to the actual world, and given that *R* is a cataclysmic type of event, while *Q* is not, the relevant possible world will contain an event of type *Q* in region *S*, but not one of type *R*. So it will be true that if there had been an event of type *P* in region *S*, it would have been followed by an event of type *Q*, and not by one of type *R*. This consequence is very counter-intuitive. Moreover, it is just one of a number of important difficulties for a possible worlds approach to subjunctive conditionals.[22]

A third objection is this: consider any possible world, *W*, which satisfies the following conditions:

---

[22] See, for example, the discussions by Bennett in 'Counterfactuals and Possible Worlds', *Canadian Journal of Philosophy*, 4, 1974, pp. 391–402, and by Jackson in 'A Causal Theory of Counterfactuals', *Australasian Journal of Philosophy*, 55, 1977, pp. 3–21.

(1) The only properties in *W* are *P*, *Q*, *F*, and *G*.

(2) There is some time when at least one individual in *W* has properties *F* and *P*, and some time when at least one individual in *W* has properties *F* and *Q*.

(3) It is true, but not a law, that everything in *W*, at any time *t*, has either property *P* or property *Q*.

(4) It is a law that for any time *t*, anything possessing properties *F* and *P* at time *t* will come to have property *G* at a slightly later time *t**.

(5) It is a law that for any time *t*, anything possessing properties *F* and *Q* at time *t* will come to have property *G* at a slightly later time *t**.

(6) No law-statements are true in *W* beyond those entailed by the previous two law-statements.

Consider now the generalized subjunctive conditional, 'For all *x*, and for any time *t*, if *x* were to have property *F* at time *t*, *x* would come to have property *G* at a slightly later time *t**.' This is surely true in *W*, on any plausible account of the truth conditions of subjunctive conditionals. For let *x* be any individual in *W* at any time *t*. If *x* has *P* at time *t*, then by virtue of the law referred to at (4), it will be true that if *x* were to have *F* at *t*, it would come to have *G* at *t**. While if *x* has *Q* at time *t*, the conditional will be true by virtue of the law referred to at (5). But given (3), *x* will, at any time *t*, have either property *P* or property *Q*. So it will be true in *W*, for any *x* whatsoever, that if *x* were to have property *F* at time *t*, it would come to have property *G* at a slightly later time *t**.

Therefore, if it were true that statements of laws are equivalent to generalized subjunctive conditionals, it would follow that it would be a law in *W* that for every *x*, and every time *t*, if *x* has *F* at *t*, then *x* will come to have *G* at a slightly later time *t**. But this law does not follow from the laws referred to at (4) and (5), and hence is excluded by condition (6). The possibility of worlds such as *W* therefore shows that law-statements cannot be analysed as generalized subjunctive conditionals. In view of this, as well as the other two reasons mentioned above, the Stalnaker/Thomason suggestion must be rejected.

2.2 LAWS AS RELATIONS AMONG UNIVERSALS

Given the failure of the above analyses of nomological statements, what account is to be offered of the nature of laws? I want to suggest that a fruitful place to begin is with the possibility of underived laws that fail to have instances of all possible types. This possibility brings the question of the nature of laws into very sharp focus, and it shows that an answer that might initially seem somewhat metaphysical is not only plausible, but virtually the only alternative.

Consider, then, the universe containing two types of particles that happen never to meet. What in that world could possibly serve as a truth-maker for some specific nomological statement concerning the interaction of particles of types X and Y? The problem is that all the states of affairs, or events, that might be taken as constituting the universe throughout all time would seem to be perfectly compatible with different, and conflicting, nomological statements concerning the interaction of these two types of particles. What, then, could possibly make it the case that one of those statements expresses a law, while the others do not?

At this point one may begin to feel the pull of some sort of non-realist view of laws. Perhaps 'law-statements' are not really statements at all, but something like inference tickets. And then, in view of the fact that in the universe envisaged there is nothing informative that one would be justified in inferring from the supposition that an X-type particle has collided with a Y-type particle, it will follow from the idea that nomological statements are nothing more than inference tickets that there cannot be, in our imaginary universe, any laws governing the interaction of particles of types X and Y.

But suppose that we resist the temptation to shift to an anti-realist conception of laws, insisting, in view of the discussion in section 2.1.1.1, that there could be underived laws governing the interaction of particles of types X and Y? Does anything of interest follow from the assumption that it is possible for there to be basic laws of a sort which fail to have instances of all possible types? In particular, can any conclusions be drawn concerning the truth-makers for the relevant nomological statements?

I want to suggest that there are two very important conclusions.

The first, and more obvious, is that the *instances* of a given law are not always sufficient to serve as truth-makers of the relevant nomological statements. In a universe in which particles of types X and Y never meet, it might be a law that when they do, an event of type P occurs. But equally, it might be a law that an event of type Q occurs. These two generalizations will not be without instances, but none of them will be such as to provide any basis for the one generalization's being nomological, and the other not. So in the case of underived laws which lack instances of all possible types, the instances cannot serve as truth-makers for the corresponding nomological statements.

But if the instances cannot serve as truth-makers, what can? Well, what are the possibilities? A natural line of thought is this: if the law in question is an underived one, then the truth-makers for the corresponding nomological statement must consist of facts concerning particles of types X and Y. One would like to be able to say, 'facts concerning the interaction of particles of types X and Y', but we have just seen that such facts do not suffice in the present sort of case. So it must be some other sort of fact concerning particles of types X and Y, and it would seem that the facts in question must somehow bear upon how particles of those types would interact if they were to collide. This in turn suggests that the facts can only be facts concerning the *powers* or *dispositions* of particles of types X and Y. Particles of type X, perhaps, are disposed to give rise to an event of type P when they are in contact with particles of type Y, and it is this dispositional fact that makes it a law that whenever particles of type X and Y collide, there is an event of type P, rather than one of type Q.

This answer, however, is unsatisfactory for a number of reasons. In the first place, in offering this sort of answer one is not really making any progress with respect to the problem of explaining nomological language in the broad sense. The question of the truth-makers for statements expressing underived laws which lack instances of all possible types has merely been replaced by that of the truth-makers of statements attributing unactualized dispositional properties to objects, and if one is willing in the latter case to say that such statements are analytically basic, and that no further account can be given of what it is for an object to have a dispositional property, one might equally well say the same thing for laws, that is, that there just are basic facts which make it the case that there are specific laws governing the behaviour of certain types of object, and no further account can be given of this. In either case one is abandoning the project of providing an account of nomological statements in non-nomological terms.

Secondly, if one offers an account of the truth-makers for such nomological statements in terms of dispositional properties of the relevant sorts of particulars, but then does not go on to offer some further account of dispositional properties, one is also abandoning the programme, outlined and defended in section 1.3, of giving a purely *extensional* account of the truth conditions of nomological statements. For dispositional predicates involve non-extensional contexts, as can be seen as follows: consider a typical sentence involving a dispositional predicate:

This is water-soluble.

There are two somewhat different ways in which such a sentence can be taken.[23] First, there is the sense in which what is being ascribed may be referred to as a 'minimal disposition'. On this first interpretation, the above sentence is analytically equivalent to:

If this is in water, then it dissolves.

where, for reasons that are familiar, the conditional cannot be construed as a material conditional.[24] Second, there is the interpretation according to which what is being asserted is that something has a minimal disposition *and* that there is some intrinsic property of the object which is the basis of that disposition. On that reading, the sentence 'This is water-soluble' will be analytically equivalent to:

This is of such a nature that if it is water, then it dissolves.

where once again the conditional cannot be a material one.

To see that the sentence involves a non-extensional context on either reading, let us suppose that while water is the only substance actually present anywhere at any time which has a freezing point of exactly 32 °F, it turns out that there is a complex compound that does not occur naturally, but which could be synthesized, and which would have a freezing point of exactly

---

[23] Mackie, *Truth, Probability, and Paradox*, Oxford, 1973, chap. 4, esp. pp. 126–33.
[24] Compare Mackie, ibid., pp. 123–5.

32 °F. As a matter of fact, however, this compound is never synthesized. This means that the following predicates are co-extensive:

   . . . is in water.
   . . . is in a substance with a freezing point of exactly 32 °F.

Suppose now that the sentence 'This is of such a nature that if it is in water, it dissolves' were extensional. The sentence that results when the predicate '. . . is in water' is replaced by the co-extensive predicate '. . . is in a substance with a freezing point of exactly 32 °F'—namely 'This is of such a nature that if it is in a substance with a freezing point of exactly 32 °F, then it dissolves'—would then have the same truth value. But it is clear that this need not follow from the fact that the predicates in question are co-extensive. If the object in question is a piece of salt, it will be soluble in water, but need not be soluble in the complex compound with a freezing point of exactly 32 °F which has never been synthesized. And as precisely the same argument applies to the more modest construal of dispositional predicates, this shows that such predicates involve non-extensional contexts. To offer an account of the truth conditions of nomological statements in dispositional terms, and then to give no account of the latter, is thus to abandon the enterprise of offering an extensional analysis of nomological statements.

The third objection to this approach can be stated more briefly. It rests upon the fact that its being a law that whenever particles of types X and Y collide an event of type P occurs cannot be adequately analysed in terms of statements asserting only that all particles of types X and Y which *happen* to exist have, *as a matter of fact*, certain dispositional properties. One would need, instead, statements which assert that it is a *law* that all particles of types X and Y have the relevant dispositional properties.

The final objection is that there are other cases that one can imagine of underived laws where there will not be any particulars that can serve as the bearers of the relevant dispositional properties. Imagine, for example, a world that, rather than containing a single law of gravitation, contains a large number of distinct laws, each of the following form:

If n is the largest integer such that the masses, m_1 and m_2, of objects a and b, are both larger than n, then the magnitude of the gravitational force which the objects exert on each other is equal to $F_n(m_1, m_2)$, divided by the square of the distance between the two objects.

Suppose further that there is no simple, general function $G(n, m_1, m_2)$ from which the different functions, F_n, can be derived by substituting the appropriate value of n—so that the different gravitational laws are not derivable from some simple, more general law. Suppose finally that the mass of the universe is equal to some finite quantity M. Then it will not, as a matter of fact, ever be the case that there are two distinct objects, each with a mass exceeding $M/2$. So for any n greater than $M/2$, there will not be any 'positive' instances for any generalizations of the form described above. Yet once again, is it not reasonable to suppose that there are laws for larger values of n, given that such laws exist for all values of n equal to or less than $M/2$?

On the dispositional approach, it *may* be possible to provide some account of possible truth-makers for some of the nomological statements with values of n in excess of $M/2$, namely, those where n does not exceed M. But it is clear that no account can be forthcoming for cases in which n exceeds M, for then even the whole universe is not large enough to be an individual possessing the relevant dispositional property.

The conclusion, then, is that an acceptable account of underived laws which fail to have instances of all possible sorts cannot be given in terms of dispositional properties possessed by particulars of the sorts whose interaction is governed by the law. But if neither facts about instances, nor dispositional facts involving the relevant sorts of particulars, can supply adequate truth-makers for law-statements of the sort we are considering, what facts about particulars can do so? It is very hard indeed to see any options that remain. As a result, I believe that we are justified in drawing the following very important conclusion: *No facts about particulars can be the truth-makers for statements expressing underived laws of the sort in question.*

But how, then, can there be such laws? The only possible answer would seem to be that it must be facts about *universals* that serve as the truth-makers for such nomological statements. But if facts about universals are the truth-makers for some law-statements, why shouldn't they be the truth-makers for *all* law-statements? This would provide one with a uniform account of the truth conditions of nomological statements, and one, moreover, that explains in a straightforward fashion the difference between generalizations which express laws, and those that are merely accidentally true.

I shall go on shortly to attempt to develop this idea in a detailed way. First, however, I should like to touch upon a very important issue which has been raised by David Armstrong, concerning the above argument for the conclusion that it is facts about universals that are the truth-makers for nomological statements.[25] The question concerns the extent to which this argument is available if one holds that there cannot be uninstantiated universals.

The following line of thought appears to lend strong support to the view that one cannot appeal to the above argument if one holds that uninstantiated universals are impossible. Consider the case of the two types of particles that never meet, and suppose that one accepts the thesis that laws are to be identified with facts about universals. What will be the truth-maker for the following statement?

It is a law that whenever an X-type particle collides with a Y-type particle, an event with property P occurs.

The most natural answer would seem to be that the truth-maker in question involves two universals standing in what may be referred to as the relation of nomic necessitation. The first of those universals will be a complex universal—the property of being a collision between an X-type particle and a Y-type particle. The second universal will just be property P. But then, given that, by hypothesis, there are never any collisions between X-type particles and Y-type particles, the first of these universals will not be instantiated. Therefore, if there are no uninstantiated universals, it will not exist at all. And so the proposed truth-maker for the nomological statement, since it consists of a certain relation between that universal and property P, will not exist either.

This line of thought is plausible enough. It rests, however, upon a particular example, and one which is, I believe, misleading in the present context. Try modifying the particle case as follows:

---

[25] Armstrong, *What is a Law of Nature?*, chap. 8, esp. pp. 119 ff.

once again, there are ten different types of fundamental particles; now, however, the behaviour of particles in interactions depends both upon the types of the interacting particles, and upon whether, at the time of the collision, they stand in relation *R*. If one considers only collisions involving two particles, there are now 110 possibilities. Suppose that 109 of these possible types of collision have been carefully studied, and 109 laws have been discovered, one for each case, which are not related in any way. The one case which has not been studied is that involving a collision of an *X*-type particle and a *Y*-type particle that do not stand in relation *R*. The reason collisions of this sort have not been studied is that while there have been very many collisions between particles of types *X* and *Y* that stood in relation *R*, there have, due to a grand accident, been none in which an *X*-type particle did not stand in relation *R* to a *Y*-type particle. Moreover, these two types of particles have now gone their separate ways, and the universe is sufficiently deterministic that *X*-type particles will never collide with *Y*-type particles at any time in the future. Thus there never has been, nor will be, a collision of an *X*-type particle with a *Y*-type particle in a situation where they do not stand in relation *R*. None the less, given that there are laws in all of the 109 other cases, it would seem very reasonable to believe that there must also be some underived law governing the interaction of *X*-type and *Y*-type particles when they do not stand in relation *R*. But since the relevant generalizations fail to have instances of the crucial sort, there will not be any extensionally characterizable facts about particulars which will determine whether it is the generalization

> Whenever there is a collision between particles of types *X* and *Y* which do not stand in relation *R*, then an event with property *P* occurs,

or the generalization

> Whenever there is a collision between particles of types *X* and *Y* which do not stand in relation *R*, then an event with property *Q* occurs,

or some other generalization which expresses the appropriate law.

The point of the modified example is that in the case of the above generalizations, the universals which exist when one of these generalizations is true, and has instances of the relevant sort, *need not differ* from those which exist when the generalization is true, but lacks such instances. The reason is that if, as suggested earlier, there are no negative universals, there will, presumably, be no complex universal corresponding to the predicate 'being a collision between particles of types X and Y which do *not* stand in relation R'. And since the universals which exist can be the same, regardless of whether the generalization has the relevant instances, we have here a case in which the view that the truth-makers for nomological statements are relations among universals, when combined with the view that there can be generalizations that fail to have instances of all possible types, and yet express laws, does not entail the existence of uninstantiated universals.

The situation, therefore, is as follows: if one rejects uninstantiated universals, while holding that relations among universals are the truth-makers for nomological statements, then it follows that there are some generalizations that cannot be viewed as expressing laws unless they have instances of all possible types. But this consequence does not obtain with respect to all generalizations— unless one holds, among other things, that there are negative universals. The possibility of laws that fail to have instances of all possible types is therefore not precluded by the view that there are no uninstantiated universals, and a defender of the latter position can appeal to the possibility of such laws in support of the claim that the truth-makers for nomological statements are facts about universals.

This line of argument does not, however, completely settle the issue raised by Armstrong. It does show that there is no inconsistency involved in accepting some uninstantiated laws, while rejecting uninstantiated universals. But this still leaves one with the question of whether these two positions are not merely mutually consistent, but rationally cotenable, and here there are at least two problems. In the first place, this combination of positions commits one to the view that while some uninstantiated laws are possible, others are not, and there is a serious question whether this can be rendered plausible. In the second place, it may be that the grounds that can be offered in support of the thesis that there cannot be any uninstantiated universals are also reasons for holding that there cannot be any uninstantiated laws.

As regards the first point, the basic problem is that the type of support that can be offered for thinking that uninstantiated laws are possible does not really differ from one possible law to another, so that it is not easy to see how one can view the conclusion as plausible in some cases but not in others. Discussion of the second point would require a rather close examination of the considerations that might be offered in support of the claim that there cannot be uninstantiated universals, to determine to what extent they also provide grounds for concluding that uninstantiated laws are also impossible. Perhaps the main point to be made, however, is one that is stressed by Armstrong— namely, that much of the appeal of an Aristotelian theory of universals arises from the fact that it can be embedded in a general, naturalistic metaphysics, according to which nothing exists beyond the spatio-temporal world. For to the extent that this is so, the view that there can be uninstantiated laws is not likely to appeal to those who reject uninstantiated universals, since uninstantiated laws, no less than uninstantiated universals, cannot be assigned location within the spatio-temporal world. And to the extent that grounds can be offered for accepting a naturalist metaphysics, there is therefore reason to reject both uninstantiated universals and uninstantiated, underived laws.

The upshot is that although it is possible, without inconsistency, to reject uninstantiated universals, while accepting some uninstantiated laws, this combination of positions may not be either very appealing, or especially plausible. So it seems likely that, in the end, the argument offered above in support of the thesis that laws are relations among universals is one that can be embraced only by those who are willing to accept a theory of universals that is at least partially Platonic.

Let us now consider how the very general idea that facts about universals can be truth-makers for nomological statements can be developed in a detailed way. Facts about universals will consist of universals having properties and standing in relations. But what are the relevant sorts of facts, and how can they serve as truth-makers for nomological statements? The basic suggestion here is that there are certain relations in which universals may stand which are such that the fact that given universals stand in one of those relations *logically necessitates* a corresponding generalization about particulars, and that when this is the case, the generalization in question expresses a law.

It is very important here that the relations in question be restricted to what may be called *genuine*, or *irreducible*, *relations among universals*. What I have in mind here is that the relations in question must satisfy the requirement that a statement asserting that such a relation holds among certain universals must not be capable of being analysed in terms of statements which refer to nothing beyond particulars, and their properties and relations. This requirement is crucial. If it were dropped, every true generalization would get classified as nomological. For suppose that everything with property P just happens to have property Q, and consider the relation R which holds between two properties A and B, just in case everything with property A also has property B. Properties P and Q will then stand in relation R, and the fact that they do trivially entails that everything with property P has property Q. So if there were no restriction to genuine, irreducible relations among universals, it would be a law that everything with property P has property Q, and the distinction between generalizations which express laws and those which are merely accidentally true would collapse.

The idea of a statement about particulars being entailed by a statement about a genuine, irreducible relation among universals is not a completely unfamiliar idea, since some philosophers have maintained, for example, that the truth-maker for a statement such as 'Nothing can be, at the same time and place, both red of shade 1 and red of shade 2' must be a relation of incompatibility between the property of being red of shade 1 and the property of being red of shade 2. In the latter case, of course, the relation would have to be a necessary one, in order for the statement about particulars to be one which expresses a necessary truth. Nomological statements, in contrast, are generally thought to be only contingently true, and if this is right, the relations among universals which are needed here will have to be contingent ones.

The idea of *contingent*, irreducible relations among universals logically necessitating corresponding statements about particulars is admittedly less familiar, and I suspect that many philosophers would feel that it is rather problematic. One very serious reason for thinking so, for example, is connected with the principle defended by Hume,[26] to the effect that there cannot be any logical

---

[26] Hume, *A Treatise of Human Nature*, bk. I, pt. III, sect. 3.

connections between distinct existences. For is not the contingent, irreducible relation among universals one state of affairs, and the corresponding general fact about particulars falling under the law a completely distinct state of affairs? And if so, how can the latter be logically necessitated by the former?

Another important objection to the present approach concerns the very idea of relations among universals which are both contingent relations and irreducible ones. Thus I suspect that many philosophers would feel that while there can be relations, such as resemblance, which are genuinely relations among universals, all of those seem to be necessary ones, and, on the other hand, while there are obviously relations, such as co-instantiation, which are contingent relations among universals, all of those appear not to be relations simply among the universals themselves.

Both objections deserve to be carefully considered. Here, however, I think it best to proceed with the exposition of the general view of the nature of laws. I shall return to these objections in section 3.1, when I consider some possible problems associated with the present account of laws.

Let us refer to properties of, and relations among, universals as *nomological* if they are contingent, irreducible properties of, or relations among, universals, and whose instantiation logically necessitates certain corresponding generalizations about particulars. How can one come to understand the meanings of terms that someone might introduce to refer to such nomological relations, or properties? If the relations or properties were observable ones, there would be no problem. But in our world, at least, the relations among universals which are the truth-makers for nomological statements appear to be unobservable. Indeed, I am inclined to go further, and to hold that in a strong sense of 'observable', such relations are necessarily unobservable. This latter claim rests, however, on some quite controversial epistemological contentions concerning non-inferential knowledge, with which I do not wish to burden myself here. And in any case, the relevant fact in the present context is simply that nomological relations among universals are not, in our world, observable. For this fact means that one is going to have to employ theoretical terms to refer to nomological relations.

Two questions, then, are crucial. First, what account is to be offered of the meaning of theoretical terms in general? Second, can that general account be applied to the special case of nomological relations? The first of these questions was discussed in section 1.2. What now needs to be considered, then, is whether the problem of specifying the meaning of terms referring to nomological relations can be handled through a straightforward application of the general approach defended there.

What is needed is a general account that can be applied to any nomological relation. However I think that the basic approach will be clearer if we begin by considering a specific nomological relation. In particular, let us consider the relation of *nomic necessitation*, where this may be informally characterized—though not *defined*—as the relation that holds between two properties, *P* and *Q*, if and only if it is a law that for all *x*, if *x* has property *P*, then *x* has property *Q*. (The presence of the undefined term 'law' precludes this from being a satisfactory definition of the expression 'nomic necessitation'.)

In order to define the theoretical expression 'relation of nomic necessitation' we need to have the appropriate analytical theory concerning this relation. That theory can be constructed as follows: first take the following two general postulates concerning the concept of a law:

(L_1): Its being the case that it is a law that *p*

logically entails

its being the case that *p*

(L_2): Its being the case that it is a law that *p*

is not logically necessary

Then apply them to the case of laws of the sort we are considering here. This gives:

(N_1): Its being the case that it is a law that for all *x*, if *x* has property *P*, then *x* has property *Q*

logically entails

its being the case that for all *x*, if *x* has property *P*, then *x* has property *Q*

(N_2): Its being the case that it is a law that for all *x*, if *x* has property *P*, then *x* has property *Q*

is not logically necessary.

The third postulate is rather more controversial. It is, however, just a thesis which has already been argued for, applied to the case of laws of the relevant type:

> (N_3): Its being the case that it is a law that for all x, if x has property P, then x has property Q
>
> is not logically equivalent to
>
> its being the case that certain facts about particulars obtain.

Now if, as I have argued, the truth-makers for law-statements are relations among universals, then its being the case that it is a law that, for all x, if x has property P, then x has property Q, is just its being the case that the universals P and Q stand in the relevant relation, i.e. the relation of nomic necessitation. When this replacement is made in (N_1), (N_2), and (N_3), and the result conjoined, one has the following analytical theory of nomic necessitation:

> (T): Its being the case that the property-universals P and Q stand in the relation of nomic necessitation
>> (1) logically entails its being the case that for all x, if x has property P, than x has property Q;
>> (2) is not logically necessary;
>> (3) is not logically equivalent to its being the case that certain facts about particulars obtain.

And given this theory, one can define the expression 'relation of nomic necessitation' using the completely general method of defining theoretical terms discussed above in section 1.2:

> Universals U and V stand in the relation of nomic necessitation means the same as
>
> There is a unique relation, K, such that K satisfies the following open formula:
>> For any property-universals P and Q, its being the case that P and Q stand in relation R
>> (1) logically entails its being the case that for all x, if x has property P, then x has property Q;
>> (2) is not logically necessary;

(3) is not logically equivalent to its being the case that certain facts about particulars obtain,'

and moreover, universals *U* and *V* stand in relation *K*.

Alternatively, one could follow Lewis in making use of identity statements in explaining theoretical terms. Then one would have:

The relation of nomic necessitation is identical with the unique relation, *K*, which satisfies the following open formula:

'For any property-universals *P* and *Q*, its being the case that *P* and *Q* stand in relation to *R*

(1) logically entails its being the case that for all *x*, if *x* has property *P*, then *x* has property *Q*;
(2) is not logically necessary;
(3) is not logically equivalent to its being the case that certain facts about particulars obtain.'

Both formulations can be condensed if one introduces the term, 'contingent', and the expression, 'genuine (or irreducible) relation among universals', understood as follows. To say that a relation is contingent is to say that its holding among any two things is never logically necessary, even given the existence of those things. To say that a relation is a genuine relation among universals is to say that its holding among any universals is not logically equivalent to the obtaining of any facts about particulars. Using these expressions, one can say:

The relation of nomic necessitation is identical with the unique contingent, genuine relation, *K*, among universals, which satisfies the following open formula:

'Its being the case that the property-universals *P* and *Q* stand in relation *R* logically entails that for all *x*, if *x* has property *P*, then *x* has property *Q*.'

Now that we have seen how to define a specific nomological relation, I should like to indicate why it seems desirable to have a *general* theory of nomological relations. The reasons will emerge, I think, if we consider an attempt to specify a small number of relations among universals which will serve as constituents in the truth-makers for all statements of laws. The view I have in mind is one that was once proposed by Armstrong, according to which only two nomological relations are needed in order to handle all non-probabilistic laws: the relation of *nomic necessitation*, and the relation of *nomic exclusion*.[27]

The reason for going beyond the relation of nomic necessitation, and introducing that of nomic exclusion, is this: consider laws expressed by statements of the following form:

It is a law that $(x)(Px \supset \sim Qx)$

where P and Q are both property-predicates. If laws of this sort were to be handled via the relation of nomic necessitation, one would be saying that the property corresponding to the predicate P stands in the relation of nomic necessitation to the property corresponding to the predicate $\sim Q$. But since Q is a property-predicate, there will be a property corresponding to the predicate $\sim Q$ only if there are negative universals. If, therefore, negative universals are rejected, another relation has to be introduced to handle laws of the above form: nomic exclusion.

Could there be law-statements whose truth-makers involved other nomological relations, or are these two relations jointly sufficient to provide truth-makers for all nomological statements? Let us consider some problematic cases. First, law-statements of the following form:

It is a law that $(x)Mx$.

Is it possible to state truth conditions for nomological statements of this sort in terms of the relations of nomic necessitation and nomic exclusion? Perhaps. A first try would be to treat its being a law that everything has property M as equivalent to its being true, of every property P, that it is a law that anything that has property P also has property M. But whether this will do depends upon certain issues about the existence of properties. If different properties would have existed if the world of particulars had been different in certain ways, the suggested analysis will not be adequate. One will have to say instead that its being a law that everything has property M is equivalent to its being a law that for every property P, anything with property P has property M—in order to exclude the possibility of there being some property Q, not possessed by any object in the world as it actually is, but which

[27] Armstrong, *Universals and Scientific Realism*, ii, pp. 157–8.

is such that if an object had property Q, it would lack property M. This revision, since it involves the occurrence of a universal quantifier ranging over properties within the scope of a nomological operator, means that laws ostensively about particulars are being analysed in terms of laws about universals. Perhaps, however, this is not unacceptable.

A rather more serious objection concerns laws expressed by statements of the form:

It is a law that $(x)(Px \supset (Qx \lor Rx))$.

If the world were partially indeterministic, there might well be laws to the effect, for example, that if an object has property P, then either it has property Q or it has property R, and yet no laws specifying *which* of those properties an object would have on any given occasion. Can the truth-conditions for nomological statements of this form be expressed in terms of the relations of nomic necessitation and nomic exclusion? The answer depends on whether there are disjunctive properties, that is, on whether, if Q and R are distinct, non-overlapping properties, there is some third property, Q or R, which is possessed by all and only those objects which either have property Q or have property R. If, as philosophers such as Armstrong and Grossman have maintained, there are no disjunctive universals, then the relations of nomic necessitation and nomic exclusion will not suffice to provide truth conditions for nomological statements of the form:

It is a law that $(x)(Px \supset (Qx \lor Rx))$.

A third case that poses difficulties concerns laws expressed by statements of the form:

It is a law that $(x)(\sim Px \supset Qx)$.

If negative universals are rejected, such laws cannot be handled in any immediate fashion by the relations of nomic necessitation and nomic exclusion. Nevertheless, this third case does not appear to raise any new issues. For if one can handle laws expressed by statements of the form 'It is a law that $(x)Mx$' in the way suggested above, one can rewrite any law-statements of the form 'It is a law that $(x)(\sim Px \supset Qx)$' in the form 'It is a law that $(x)(Px \lor Qx)$', and then apply the method of analysis suggested for laws expressed by statements of the form 'It is a law that $(x)Mx$.' The result will be a

law-statement that is conditional in form, with a positive antecedent and a disjunctive consequent, which is the case just
considered.

The conclusion seems to be this. The relation of nomic
necessitation by itself does not provide a satisfactory account
unless there are both negative and disjunctive properties. Supplementing it with the relation of nomic exclusion may allow one
to dispense with negative properties, but not with disjunctive ones.
So if neither negative nor disjunctive universals can exist, then
there may have to be nomological relations other than nomic
necessitation and nomic exclusion in order for there to be truth-
makers for all statements of non-probabilistic laws.

It is possible, however, to establish a much stronger and more
general conclusion, concerning the nomological relations that may
be needed to provide truth-makers for all statements of non-
probabilistic laws. The argument rests upon what appears to be a
very plausible principle, which Armstrong refers to as the
Principle of Instantial Invariance: 'For all n, if a universal is n-adic
with respect to a particular instantiation, then it is n-adic with
respect to all its instantiations (it is n-adic *simpliciter*).[28]

If this principle is correct, then every relation has some number,
k, associated with it, where k is the number of individuals involved
in any instantiation of that relation. If, then, S is any finite set of
nomological relations, there will have to be some number, n, such
that n is the largest number associated with any of the relations in
the set S. It is logically possible, however, that our world contains
some underived law that could be expressed by a statement of the
form:

It is a law that $(x)(Px \supset (Q_1x \vee \ldots \vee Q_nx))$

where P and all of the Q_i are property-predicates, none of which
are associated with overlapping properties. If this statement is to
express an *underived* law, the truth-maker for it will have to be a
single nomological relation which relates all of the properties
associated with the predicates contained in the statement. But
there are $(n + 1)$ such properties. Therefore none of the
nomological relations in S can serve to relate them. The

---

[28] Armstrong, ibid., p. 94. Armstrong argues that this principle follows from the
fact that a universal must, by definition, be strictly identical in its different
instantiations.

conclusion, then, is that there is no finite set of nomological relations which will suffice to provide truth-makers for all possible nomological statements.

Let us now turn to the problem of offering an account of nomological relations in general. There are three further ideas that are needed in order to set out such an account. The first is that of the *universals associated with a proposition.* (Alternatively, as we shall see, one can speak of the universals involved in a *sentence.*) A full explication of this concept would require a lengthy detour through some difficult issues in semantics and philosophy of mind. For present purposes, however, I think that a brief, informal characterization should suffice. So let *p* be any proposition, and assume that one has a language containing some sentence, *S*, which expresses the proposition that *p*. Assume further that it is possible to find, or to introduce, a set of primitive predicates such that every predicate in sentence *S* can be analysed in terms of those primitive predicates, and where a predicate is primitive if and only if it satisfies the following condition: Each primitive predicate must have a *single* universal associated with it. Let *S** be any sentence that results from *S* via analytical transformations which eliminate all non-primitive predicates in *S* in favour of primitive ones. One can then offer the following explanation of what is meant by the expression 'the universals involved in a proposition':

Universals *U* is involved in proposition *p*

means the same as

For all sentences *S* and *S**, if *S* expresses proposition *p*, and *S** is a translation of *S* into primitive vocabulary, then there is some predicate *P* in *S** such that *U* is associated with *P*.

Alternatively, if one wishes to avoid reference to propositions, one can instead talk about the universals associated with a sentence, understood as follows:

Universal *U* is associated with sentence *S*

means the same as

For any sentence *S**, if *S** is a translation of *S* into primitive vocabulary, then there is some predicate *P* in *S** such that *U* is associated with *P*.

The exact relation between these two approaches will depend, of course, upon the interpretation that is assigned to the crucial term 'translation'.

The second, and related idea is that of the logical form or structure of a proposition. It is natural to view this form as specified by a *construction function* which maps ordered n-tuples of universals into propositions. Thus one could have, for example, a construction function, K, such that the value of K when applied to the ordered couple (redness, roundness) is identical with the proposition that all red things are round.

Conceived in the most general way, some construction functions will map ordered n-tuples of universals into propositions that involve as constituents universals not contained in the original n-tuple. Thus G could be a function so defined that G (property P) is identical with the proposition that everything with property P is green.

Other construction functions will map ordered n-tuples of universals into propositions that do not contain, as constituents, all of the universals belonging to the n-tuple. H could be a function so defined that H(property P, property Q) is identical with the proposition that everything has property Q.

In order to capture the notion of logical form, one needs a narrower notion of construction function, namely, one in which something is a construction function if and only if it is a mapping from ordered n-tuples of universals into propositions that contain, as constituents, all and only those universals belonging to the ordered n-tuple. In this narrower sense, K is a construction function, but G and H are not.

The final idea required is that of a *universal being irreducibly of order m*. Properties of, and relations among, particulars are universals of order 1. If nominalism is false, they are irreducibly so. A universal is of order 2 if it is a property of universals of order 1, or a relation among things, some of which are universals of order 1, and all of which are either universals of order 1 or particulars. It is irreducibly so if it cannot be analysed in terms of universals of order 1. And in general, a universal is of order $(m + 1)$ if it is a property of universals of order m, or a relation among things, some of which are universals of order m, and all of which are either particulars or universals of order m or less. It is irreducibly of order $(m + 1)$ if it cannot be analysed

in terms of particulars and universals of order *m* or less.
With these concepts, it is a straightforward matter to construct a
theory encompassing all possible non-probabilistic nomological
relations. Again, one starts from the following two general
postulates:

(L_1): Its being the case that it is a law that p

logically entails

its being the case that p

(L_2): Its being the case that it is a law that p

is not logically necessary.

When we were dealing with the specific relation of nomic
necessitation, the next step was to replace (L_1) and (L_2) by (N_1)
and (N_2), where the latter contained a specification of the logical
form possessed by law-statements whose truth-makers involved
the relation of nomic necessitation. A similar move is necessary at
this point:

(T_1): Its being the case that it is a law that $K(P_1, P_2, \ldots P_n)$

logically entails

its being the case that $K(P_1, P_2, \ldots P_n)$ where K is
some construction function whose values are universally
quantified propositions, and $P_1, P_2, \ldots P_n$ are
universals of the appropriate types.

(T_2): Its being the case that it is a law that $K(P_1, P_2, \ldots P_n)$

is not logically necessary.

Next, we need an irreducibility postulate—comparable to (N_3)
in the theory of nomic necessitation. If we were interested only in
first-order laws—that is, in laws expressed by generalizations all of
whose quantifiers range over particulars—then the following
would be what was wanted:

Its being the case that it is a law that $K(P_1, P_2, \ldots P_n)$

is not logically equivalent to

Its being the case that certain facts about particulars
obtain.

However, one would like a theory that also covers the possibility of higher-order laws—that is, laws expressed by generalizations some of whose quantifiers range over universals. Therefore, rather than the above postulate, what is needed is the following slightly more complicated one:

(T_3): Its being the case that it is a law that $K(P_1, P_2, \ldots P_n)$ is not logically equivalent to

its being the case that certain facts about individuals of level m or below obtain, where m is the level of the highest order individuals quantified over in the proposition $K(P_1, P_2, \ldots P_n)$.

The next step involves transforming postulates (T_1), (T_2), and (T_3), first, by appealing to the idea that the truth-makers for nomological statements are relations among universals, and second, by individuating each nomological relation by reference to the logical form of the corresponding non-nomological proposition, the form in turn being specified by the relevant construction function, K. When these transformations are carried out on (T_1), (T_2), and (T_3), and the results conjoined, we have the following general theory:

(T): Its being the case that universals $P_1, P_2, \ldots P_n$ stand in the nomological relation, R_K

(1) logically entails the universally quantified proposition, $K(P_1, P_2, \ldots P_n)$;

(2) is not logically necessary;

(3) is not logically equivalent to its being the case that certain facts about individuals of level m or below obtain, where m is the level of the highest order individuals quantified over in the proposition $K(P_1, P_2, \ldots P_n)$.

By substituting different construction functions in this general theory, one can generate an analytical theory of any specific nomological relation. What I am interested in here, however, is the general concept of a nomological relation. The basic analysis, therefore, will be simply this:

N is a nomological relation

means the same as

There is some construction function, K, whose values are universally quantified propositions, such that N is identical with the unique relation, R, which satisfies the following open sentence:

For all universals, $P_1, P_2, \ldots P_n$, its being the case that $P_1, P_2, \ldots P_n$ stand in relation R

(1) logically entails its being the case that $K(P_1, P_2, \ldots P_n)$;
(2) is not logically necessary;
(3) is not logically equivalent to its being the case that certain facts about individuals of level m or below obtain, where m is the level of the highest order individuals quantified over in the proposition $K(P_1, P_2, \ldots P_n)$.

This account can, however, be condensed, and rendered more perspicuous, by utilizing the concept of a contingent relation, and that of a relation's being irreducibly of order m. The following is a natural restatement:

N is a *nomological relation*

means the same as

N satisfies the following conditions:
(1) N is an n-ary relation among universals.
(2) N is irreducibly of order $(m + 1)$, where m is the level of the highest order element that can enter into relation N.
(3) N is a contingent relation among universals.
(4) There is a construction function, K, such that

(a) K generates universally quantified propositions of order $(m-1)$;
(b) its being the case that $P_1, P_2, \ldots P_n$ stand in relation N logically entails its being the case that $K(P_1, P_2, \ldots P_n)$, and there is no other relation, R, of which this is true.

This characterization of the theoretical concept of a nomological relation can in turn be used to formulate truth conditions for nomological statements. The place to begin, it would seem, is with the notion of a *basic*, or underived nomological generalization, where this is a generalization that is true by virtue of a *single* nomological relation that obtains among *all and only* those universals associated with predicates involved in the generalization. This concept can be defined as follows:

> G is a basic, nomologically true generalization
>
> means the same as
>
> There is a proposition, p, which is expressed by G and there exists a nomological relation, N, and an associated construction function, K, and universals P_1, P_2, ... P_n such that
>
> (1) it is not logically necessary that p
> (2) the proposition, p, is identical with the value of $K(P_1, P_2, ... P_n)$;
> (3) it is true that P_1, P_2, ... P_n stand in relation N;
> (4) the proposition that P_1, P_2, ... P_n stand in relation N logically entails the proposition, p.

Can the class of basic nomological generalizations be identified with the class of generalizations that express laws? It would seem not. For suppose that it is a law that everything with property P has property Q, and also a law that everything with property Q has property R. Then it must also be a law that everything with property P has property R. But this could be the case even if properties P and R did not themselves stand in the relation of nomic necessitation; it would suffice that that relation obtains between properties P and Q, and between Q and R. In such a case it would be law that everything with property P has property R, but the generalization in question would not qualify as a basic nomological one. Thus, while every generalization that expresses a basic nomological truth necessarily expresses a law, the converse is not the case.

A natural response to this problem is to define a wider notion of a nomological generalization, either basic or derived, along the following lines:

> H is a nomologically true generalization
>
> means the same as
>
> H is a generalization, and there are generalizations G_1, G_2, ... G_n, each of which is a basic, nomologically true generalization, such that H is entailed by the conjunction of G_1, G_2, ... G_n.

It appears, however, that the class of nomological generalizations, thus defined, is too wide to be identified with the class of generalizations that express laws. For suppose that it is a nomologically true generalization that all salt, when in water, dissolves. Then on the above account it would also be a nomologically true generalization that all salt, when it is both in water and in the vicinity of a piece of gold, dissolves. This is not, however, a generalization that one naturally views as expressing a law. Moreover, this feeling is reinforced if one considers the relation between laws and corresponding counterfactuals. If it were a law that all salt, when both in water and in the vicinity of a piece of gold, dissolves, then it would seem that that should license counterfactuals such as 'If this piece of salt were in water, and were not dissolving, then it would not be in the vicinity of a piece of gold.' As this counterfactual seems unacceptable, it appears that not all nomologically true generalizations express laws.

How is a distinction to be drawn within the class of nomological statements in the broad sense between those that express laws and those that do not? This is one of the problems discussed in great detail by Hans Reichenbach in his important but much neglected book, *Nomological Statements and Admissible Operations.*[29] As far as I am aware, the only satisfactory solution to the problem is along the lines developed by Reichenbach.[30] That solution is, however, quite complex, and I do not believe that it would be especially useful to go into it here. Interested readers can turn to Salmon's essay, and thence to Reichenbach's own discussion.

There is one final point that needs to be noted. I indicated earlier that it is crucial to the present approach that a nomological relation be genuinely a relation among universals, and nothing else—as contrasted, for example, with a relation that is apparently among universals, but which can be analysed in terms of properties of, and relations among, particulars. Hence the requirement—condition (2)—that a relation, to be nomological, always be

[29] Reichenbach, *Nomological Statements and Admissible Operations*, Amsterdam, 1954. Reissued as *Laws, Modalities, and Counterfactuals*, Berkeley and Los Angeles, 1976, with a very helpful foreword by Wesley Salmon. Salmon's introductory discussion is also available as 'Laws, Modalities, and Counterfactuals' in *Synthese*, 35, 1977, pp. 191–229.

[30] I am indebted to Stephen C. Hetherington, in 'Tooley's Theory of Laws of Nature', *Canadian Journal of Philosophy*, 13/1, 1983, pp. 101–5, for pointing out difficulties in a simpler account that I proposed in an earlier discussion.

irreducibly of a level greater than the level of the highest order element which can enter into the relation. If this requirement were not imposed, every true generalization would get classified as nomological, as was shown above.

But while condition (2) is essential, it is not quite adequate. For suppose that it is a law that everything with property S has property T, and that the truth-maker for this law-statement is the fact that S and T stand in a certain relation W, where W is irreducibly of order 2. Then one can define a relation R as follows: properties P and Q stand in relation R if and only if everything with property P has property Q, and properties S and T stand in relation W. So defined, relation R will not be analysable in terms of universals of order 1, so condition (2) will not be violated. But if relations such as R were admitted as nomological, then, provided that there was at least one true nomological statement, all generalizations about particulars would get classified as nomological.

There are alternative ways of coping with this difficulty. One is to replace condition (2) by:

(2*) If $R(U_1, U_2, \ldots U_n)$ is analytically equivalent to C_1 & C_2 & $\ldots C_m$, then every nonredundant C_i—that is, every C_i not entailed by the remainder of the conjunctive formula—is irreducibly of order $(m + 1)$.

This condition blocks the above counterexample. But given the somewhat *ad hoc* way in which it does so, one might wonder whether there may not be related counterexamples which it fails to exclude. What is one to say, for example, about a relation R defined as follows: properties P and Q stand in relation R if and only if either everything with property P has property Q, or properties P and Q stand in relation W? I would hold that this is not a counterexample, on the ground that there cannot be disjunctive universals. However one might prefer an account which was ontologically more neutral. One possibility would be to replace condition (2) by:

(2**) Relation N is not analysable in terms of other universals of *any* order.

I am inclined to believe that this somewhat more radical approach is probably preferable. The issue is, however, complex, and I do not think it is important to explore it here.

2.3 MORE COMPLEX LAWS

The above account of the truth-makers for nomological general-izations is straightforward in the case of statements such as:

Whenever a particle has spin S and charge C, it also has mass M;

or

Whenever a substance has molecular structure T, and is in water, it dissolves.

The situation is less straightforward, however, in the case of more complex laws. Consider, for example, a world in which Newtonian physics is true. What would be the truth-maker for the statement:

It is a law that for all x and all y, the gravitational force which x exerts on y is equal in magnitude to a constant, k, times the product of the mass of x and the mass of y, divided by the square of the distance between x and y?

The answer for this sort of case is rather more difficult. The problem is that it is unclear precisely what universals are constituents of the relevant truth-makers. One way of seeing this is by considering the more specific law-statements that are entailed by the statement[31] expressing the general Newtonian law of gravitation. It entails, for example the following nomological statement:

It is a law that, for all x and all y, if the mass of x is one unit, and the mass of y is one unit, and the distance between x and y is one unit, then the gravitational force which x exerts on y is equal in magnitude to the gravitational constant, k.

Now if this latter statement expressed an *underived* law, it would be easy to specify plausible truth-makers for it. For if one can maintain, as seems plausible, that objects that have a mass of one unit share a certain property, and similarly, that pairs of objects

---

[31] The above statement is not, of course, a full statement of the Newtonian law of gravitation, since it specifies the magnitude, but not the direction, of the gravitational force. In our present discussion, however, it will suffice to consider only statements dealing with the magnitude of the gravitational force.

that are one unit apart, or that exert a gravitational force of k units upon each other, share certain relations, then one can identify the truth-maker for the specific law with the appropriate nomological relation among those universals. But what if the law in question is a derived one—as it will be in a Newtonian universe? Will the nomological statement then have a different truth-maker? Or will it have the same truth-maker? And if so, how will that truth-maker be related to the state of affairs that serves as the truth-maker for the statement expressing the general Newtonian law of gravitation?

There are two main views that might be adopted here. The first is that the relation among universals that will be the truth-maker for the specific nomological statement in a world where it expresses an underived law will also be the truth-maker in the fully Newtonian world, and that in a world where Newton's law of gravitation obtains, the truth-maker for any statement expressing that general law will consist of nothing over and above the sum of the states of affairs that are the truth-makers for the more specific nomological statements which it entails.

The second view is that the sum of the truth-makers for the more specific nomological statements does not constitute an adequate truth-maker for any statement that expresses the general Newtonian law of gravitation: some distinct truth-maker is required. Moreover, given this distinct truth-maker, there will be no need for individual truth-makers for the more specific nomological statements which follow from a statement of the Newtonian law of gravitation.

Which of these views is more plausible? One reason that might be offered for preferring the second alternative is this: it is natural to think of the Newtonian law of gravitation as a single law; on the first view, however, what one has is not a single law, but a possibly infinite—and perhaps even non-denumerable—set of laws, one corresponding to at least each ordered triple of the form (m_1, m_2, d), where m_1 and m_2 are instantiated mass-properties, and d an instantiated distance-relation. On the second view, in contrast, there is a single law, of what might be referred to as a higher-order type, since it would seem to involve not only quantification over ordinary, first-order particulars, but also quantification either over numbers, or over properties and relations of the relevant types. The second view, then, seems more natural. It is not clear, however, how much weight should be assigned to this consideration.

A second possible objection to the first view is this: consider a world which is Newtonian, and where the total mass is equal to some finite quantity, m_0: given that the Newtonian law of gravitation obtains, there will be derived laws concerning the gravitational force of attraction between bodies whose masses are greater than m_0. But if the first view is adopted, such laws may not be possible. In particular, if there cannot be any uninstantiated universals, then in the universe in question there will not be any property of having mass m, if m is greater than m_0, and therefore the relation between universals which serves, on the first view, as the truth-maker for the derived nomological statement in question, will not exist. That statement cannot, therefore, express a law in the universe being considered.

This objection turns upon the crucial metaphysical claim that there cannot be uninstantiated universals. If that contention is correct, then this second objection does provide a strong argument in support of the second view. Earlier, however, I argued that one should admit the possibility of underived laws which lack crucial instances, and we shall see in section 3.1 that if this possibility is accepted, it entails, in some cases, the possibility of uninstantiated universals, allowing which the present objection will be undercut.

The third objection is this: suppose that the first view is right, so that there is no distinct truth-maker for any statement expressing the Newtonian law of gravitation: it is true simply by virtue of the truth-makers for all of the specific, derived nomological statements. In order for the Newtonian law of gravitation to obtain, then, there has to be a perfect agreement among the truth-makers for each of the (possibly infinite) set of nomological statements expressing the relevant derived laws, for otherwise not all of the laws dealing with the gravitational force which one body exerts on another would fall under a single, overarching, higher-order law. But is this really probable? Would not such an outcome be most remarkable indeed?

In contrast, if the second view is adopted, there is a single distinct truth-maker for any statement expressing the Newtonian law of gravitation, and this higher-order truth-maker also serves as a truth-maker for all of the more specific law-statements which are entailed by that statement. There is therefore no grand accident that stands in need of explanation.

This third objection provides, I believe, a strong reason for accepting the second view of more complex laws. For as David Armstrong has pointed out, this objection does underline the peculiarity of the first view, according to which regularities are explained—by relations among universals—at the first level, but left unexplained at the second level.

Let us ask, then, what universals could be involved in a state of affairs which might serve as a truth-maker for some statement expressing the (non-vector part of the) Newtonian law of gravitation. One can begin by considering the formulation offered above:

> It is a law that for all *x* and all *y*, the gravitational force which *x* exerts on *y* is equal in magnitude to a constant, *k*, times the product of the mass of *x* and the mass of *y*, divided by the square of the distance between *x* and *y*.

If the predicates involved in this statement had single universals associated with them, what sorts of universals would they be? Among the non-mathematical predicates contained in the above statement is the term 'mass'. This term stands for a function which associates with any first-order particular some number representing its mass. What that number is depends, however, not only upon the particular object, but also upon the unit of mass which has been chosen. It would seem, then, that if there is a single universal associated with the term 'mass', it must be a three-termed relation holding among first-order particulars, a unit of mass, and numbers. Similar remarks apply to 'gravitational force', and to 'distance between'. If the latter, for example, has a single universal associated with it, it would seem that it will have to be a four-termed relation holding among pairs of first-order particulars, a unit of length, and numbers.

Can one then say that the state of affairs consisting of the holding of a certain nomological relation among these universals is the truth-maker for the statement expressing the Newtonian law of gravitation? I want to suggest that there are important considerations that tell against this view. In the first place, there is a principle advanced by Armstrong in his discussion of universals, which he refers to as the Weak Principle of Order Invariance:

If there are a number of particulars, and a relation which relates them, then each particular is a particular of the same order.[32]

This principle, in contrast to the Strong Principle of Order Invariance, does not assert that if $x_1, x_2, \ldots x_i, \ldots x_n$ stand in relation R, and similarly for $y_1, y_2, \ldots y_j, \ldots y_n$, that x_i and y_j must be particulars of the same order. What it asserts is that all of the x_i must be particulars of the same order, and similarly for all of the y_j. Armstrong suggests that while the Strong Principle appears dubious, the Weak Principle seems sound. If this view is correct— and it strikes me as a plausible one—then it would seem that there cannot be universals of the sort needed for the account just outlined. For whatever numbers are—whether classes of classes, or properties of classes, or properties of properties, etc.—it seems most unlikely that they are to be viewed as first-order particulars, and so there cannot, for example, be a relation which holds among first-order particulars, a unit of mass, and numbers.

A second problem concerns the necessary reference to a unit— of mass, or of length, and so on. How are such units to be specified? There appear to be two main options here. One is to specify the unit by reference to some (first-order) particular that serves as the standard: a rod is a metre long if it is precisely equal in length to the standard metre. But if this approach is adopted, the law-statements will incorporate reference to particular individuals—something which is surely to be avoided if at all possible.

The other main alternative is to specify the unit by reference to some type of particular. A unit of mass might be defined, for example, in terms of the mass of a certain number of neutrons. But if this were done, it would follow that the Newtonian law of gravitation requires the existence of neutrons—a conclusion that is also unpalatable.

A third objection to the above approach is that it commits one to having numbers in one's ontology. This objection will not seem very impressive if one finds the existence of numbers completely unproblematic, though even then it seems to me that, other things being equal, an account that avoids such ontological commitment should be viewed as preferable.

[32] Armstrong, *Universals and Scientific Realism*, ii, p. 142.

There is a final, and related point, which is much more serions. It is that even if there are numbers, it is really very puzzling why nomological statements need incorporate any reference to them, since, as Hartry Field emphasizes in his recent, and very stimulating book, *Science Without Numbers*, numbers are entities that are completely extrinsic to the processes that fall under causal laws:

> But even on the platonistic assumption that there are numbers, no one thinks that those numbers are causally relevant to the physical phenomena: numbers are supposed to be entities somewhere outside of space-time, causally isolated from everything we can observe. If, as at first blush appears to be the case, we need to invoke some real numbers like 6.67×10^{-11} (the gravitational constant in $m^3/kg^{-1}/s^{-2}$) in our explanation of why the moon follows the path that it does, it isn't because we think that that real number plays a role as a *cause* of the moon's moving that way; it plays a very different role in the explanation than electrons play in the explanation of the workings of electric devices. The role it plays is as an entity *extrinsic to the process to be explained*, an entity related to the process to be explained only by function (a rather arbitrarily chosen function at that). Surely then it would be illuminating if we could show that a purely intrinsic explanation of the process was possible, an explanation that did not invoke functions to extrinsic and causally irrelevant entities.[33]

Field goes on to propose what he suggests is a plausible methodological principle, to the effect that '*underlying every good extrinsic explanation there is an intrinsic explanation*'.[34] I want to suggest that a parallel principle is also very plausible in the case of laws, namely, that corresponding to every extrinsic formulation of a law—that is, every formulation of a law which involves reference to entities that are nomologically unconnected with events or states of affairs falling under the law—there must be some purely intrinsic formulation of the law, which is free of all such reference. And if this principle is sound, the present approach to the problem of specifying truth-makers for complex nomological statements must be rejected as unsatisfactory.

What, then, is the alternative? The most promising approach, I believe, involves reformulating the law-statements in question along the general lines developed by Field in *Science Without*

---

[33] Field, *Science Without Numbers*, Princeton, 1980, p. 43.
[34] Ibid., p. 44.

Numbers. He attempts to make plausible the claim that all of physics can be formulated without reference to numbers by showing, in some detail, how such a programme can be carried out in the case of Newtonian gravitational theory. For our purposes here, however, it will suffice to take a brief look, first at Field's treatment of Newtonian space-time, and second at his treatment of quantities.

Field's treatment of Newtonian space-time is modelled on Hilbert's approach to Euclidean geometry. Hilbert showed that although Euclidean geometry is concerned with length of lines and size of angles, it is possible to set out a satisfactory system of axioms for Euclidean geometry which does not involve any reference to numbers.[35] In the case of Newtonian space-time, a different geometry is needed, and Field develops his account by building on the intrinsic treatment of affine geometry set out by Szczerba and Tarski.[36] What is of interest to us here is the primitive predicates that are required. It turns out that only three primitive predicates are needed for the description of the structure of Newtonian space-time. These are predicates that stand for three relations among space-time points. First, there is the three-place relation of *betweenness*. Secondly, the two-place relation of *simultaneity*. Finally, the four-place relation of *spatial congruence*, where this is the relation that holds between the segment defined by two simultaneous space-time points, x and y, and the segment defined by two other (mutually) simultaneous space-time points, z and w, when and only when the segments are equal in magnitude. No other primitives are required beyond those associated with these three relations.

Before discussing the significance of this result in the present context, let us consider Field's treatment of quantities. His basic approach, as he emphasizes, is essentially that developed in the area known as measurement theory. Within this general approach, there are slightly different alternatives, with respect, among other things, to the primitive predicates employed. The approach employed by Field can, for our purposes, be viewed as involving two primitives for each type of quantity, where each primitive stands for a relation among space-time points. One of the

[35] Hilbert, *Foundations of Geometry*, La Salle, Ill., 1971.
[36] Szczerba and Tarski, 'Metamathematical Properties of Some Affine Geometries', in *Logic, Methodology, and Philosophy of Science*, ed. Bar-Hillel, Amsterdam, 1965, pp. 166–78.

primitives will, depending upon the type of quantity in question, either stand for the three-place relation of *betweenness* with respect to the quantity in question, or else stand for the two-place relation of being *less than* with respect to the quantity in question. In the case of mass, for example, one would have a primitive predicate standing for the relation which obtains when one space-time point has less mass than another. Secondly, there will be a primitive that stands for the relation of *congruence* with respect to the quantity in question. In the case of mass, this will be the relation that obtains among x, y, z, and w when and only when, to express it intuitively, the difference in mass between x and y is equal in absolute value to the difference in mass between z and w.

There are metaphysical objections that can be raised to the above account, but before turning to them, it is important to grasp clearly what the above account *does* accomplish, namely, it enables one to avoid the four objections that confronted the previous account. In the first place, none of the relations introduced either in the account of space-time or in the account of quantities ever obtains among different sorts of object. Indeed, the only objects that ever enter into these relations are space-time points. So the Weak Principle of Order Invariance is certainly satisfied. Secondly, there is no reference to numbers, so the third and fourth objections drop away. And finally, since there is no reference to numbers, neither is there any need to specify units, so the second objection is also defused.

These are very substantial gains. There are, however, at least three features of the account that seem unhappy. The first concerns the objects that enter into the relations employed in the treatment of quantities. On Hartry Field's account, those objects are space-time points. One objection to this would be that space-time points are either fictions, or, alternatively, entities which are logical constructions out of more basic entities. This objection raises a difficult issue. Fortunately, I do not believe that we need pursue it here. If one has doubts, either about the existence of space-time points, or about whether they are ontologically basic, one can easily recast the above treatment in terms of physical objects and parts thereof.

A second problem is this: the relation of spatial congruence which is introduced to describe the structure of space-time appears to be a simple and natural relation. One needs to compare lengths, and each length is naturally specified by two space-time points. Hence the four-place relation of spatial congruence. In contrast, the four-place congruence relations introduced to deal with quantities, such as mass, seem distinctly artificial, and one might easily come to feel that they may have been introduced simply because they were necessary to achieve a certain technical result.

The view that there is something artificial about the relations of congruence with respect to different sorts of quantity may be right. This issue is not, however, a crucial one, since it is possible to introduce other, simpler and more natural relations, in terms of which the congruence relations can be defined. In particular, they can be defined in terms of the part/whole relation together with the two-place relation of being equal with respect to a certain quantity. Consider, for example, the four-place relation of congruence with respect to mass. To say that this holds among x, y, z, and w is to say that the difference in mass between x and y is equal in absolute value to the difference in mass between z and w, and if one leaves aside cases in which there is overlap between some of x, y, z, and w, this is logically equivalent to saying that either the whole composed of x together with z is equal in mass to the whole composed of y together with w, or the whole composed of x together with w is equal in mass to the whole composed of y together with z. Any artificiality is, therefore, easily eliminated.

This brings me to the third, and by far the most serious problem. Consider the relation of being heavier than. Even if one treats this as a relation between physical objects, doesn't it seem that there is something unsatisfactory about introducing this as a primitive relation on a par with the relation of spatial betweenness? For while whether one object is between two others does not depend upon the intrinsic properties of the three objects, whether one object is heavier than another does surely depend upon the intrinsic properties of the two objects. An account, then, that fails to bring in any reference to those properties is ontologically misleading.

What sort of dependence is involved here? There seem to be two alternatives. One is that the dependence is a *logical* one. Given, for example, that x has some mass property, m_1, and y has some distinct mass property, m_2, then either it logically follows that x is greater in mass than y, or else it logically follows that y is greater in mass than x.

On this first view, then, the betweenness, less than, and congruence relations that have been introduced to deal with quantities are *internal* relations. But if this is in fact the case, then it would seem that nomological relations among such universals could not serve as the truth-makers for statements expressing more complex laws such as Newton's law of gravitation. The reason is that it seems plausible to accept what Armstrong refers to as the Reductive Principle for internal relations: 'If two or more particulars are internally related, then the relation is nothing more than the possession by the particulars of the properties which necessitate the relation.'[37]

The other alternative is that the dependence is *nomological*, rather than logical. On this view, a relation such as 'having more mass than' would be an external relation, and there would be laws such that, given that x has some mass property, m_1, and y some mass property, m_2, it would be nomologically determined whether x is greater in mass than y.

This alternative appears implausible however. In the first place, it seems to involve an unnecessary multiplication of universals. And secondly, if the relation of having more mass than is an external relation that is only nomologically connected with the mass-properties of objects, it must be possible for there to be a world where the connection in question does not obtain, assuming that laws of nature are contingent, rather than necessary. This means that it must be possible, for example, for there to be a world containing objects A, B, and C, such that A has more mass than B, and B more mass than C, but C more mass than A. Yet is it not reasonable to hold that the relation of having more mass than is necessarily transitive? If so, then it would seem that the second alternative must be rejected, since it can provide no account of how this could be so.

But if the relations of betweenness, less than, and congruence cannot be external relations between objects, and if, when viewed as internal relations, they cannot enter into the relevant truth-makers for the more complex laws, are we not forced to conclude that this whole approach must be abandoned? The answer is that there is one alternative that has not yet been considered—namely, the view that the relations in question are relations not between

---

[37] Armstrong, *Universals and Scientific Realism*, ii, p. 86.

space-time points or physical objects, but between the properties of such things.

This alternative is, in fact, mentioned by Field in his discussion:

A possible approach to a coordinate-independent treatment of, say, temperature, would be to introduce a continuum of temperature properties, each one the property of having such and such specific temperature. One could then describe the structure of that system of properties not via numbers, but via certain intrinsic relations among them, say, the relations of betweenness and congruence . . .[38]

Field does not himself adopt this approach. His reason, however, is not that this alternative is technically flawed. It is merely that he prefers as nominalistic a treatment as possible.

If this alternative is adopted, the relations in question will be second-order ones. How can they be characterized? The best way of approaching this question, I think, is by considering related second-order properties. In particular, let us focus upon the case of mass, and consider the second-order property of being a mass-property —that is, the property that a first order property has if and only if it is a property that endows objects with mass. What is it for some first-order property, P, to be a mass-property? I want to suggest that it is a matter of its entering into a certain sort of law. There are, however, a number of slightly different possibilities with respect to the relevant sort of law. One suggestion is that P can be a mass-property only if the following is the case:

It is a law that, for all x and y, if x and y both have property P, and if they are the only things that exist, then, if they collide, they will undergo accelerations that are equal in magnitude.

Alternatively, if one has the notion of causation at one's disposal, one might prefer to say that P can be a mass-property only if the following is the case:

It is a law that, for all x and y, if x and y both have property P, and if they collide, and are not causally acted upon at that time by any other objects, then they will undergo accelerations that are equal in magnitude.

These differences are probably not crucial. What is important is a feature that Armstrong has pointed out—namely, that the laws

[38] Field, op. cit., p. 55.

appealed to here appear to be uninstantiated ones. This is especially clear in the case of the first sort of law, with its reference to the counterfactual situation where x and y are the only things that exist, but given the apparent nature of our world, it also seems most unlikely that there have ever been two objects that were, at some time, causally isolated from all other objects—which means that laws of the second sort are, in all probability, uninstantiated as well.

Is this appeal to laws that are probably uninstantiated an unavoidable feature of the present approach? I think that, unfortunately, it probably is. A different approach will be called for, then, if one shares Armstrong's doubts concerning the possibility of uninstantiated laws.

The situation is similar, though slightly more complicated, in the case of relations between mass properties. Consider the relation that holds between two mass-properties when and only when the one endows something with less mass than the other. A first attempt at capturing that relation would be to say that two mass properties, P and Q, stand in that relation if and only if the following expresses a law:

> For all x and y, if x has property P and y has property Q, and x and y are the only things that exist, then, if x and y collide, x will undergo greater acceleration than y.

But this is not entirely satisfactory, as can be seen from the fact that while this relation is irreflexive and asymmetric, it need not be transitive. The relation thus defined can be used, however, to capture the relation that one wants. Let us say that two mass properties, P and Q, stand in relation R if the above expresses a law. Then P will be a property that endows objects with less mass than property Q if and only if the following is the case:

> P and Q are mass properties standing in relation R, and it is a law that for all properties S, T, and U, if S and T stand in relation R, and T and U stand in relation R, then so do S and U.

The relation thus defined will be transitive, as well as irreflexive and asymmetric.

Armstrong has questioned, however, whether the above approach succeeds in capturing everything that one normally takes to be involved in the concept of mass. For example, is it not a necessary truth that if the mass of A is equal to the mass of a part of B, then the mass of A cannot be greater than the mass of B? Yet this does not follow from the analyses just offered.

Armstrong's objection seems sound; however I believe that the above analyses can be modified to deal with this point. For instead of viewing the above accounts as explaining what it is to be a mass-property, and what it is for one property to endow things with less mass than another property does, one can view them as explaining certain corresponding, more elementary, 'proto-mass' concepts. One can then offer the following accounts of what it is for something, first, to be a mass-property, and secondly, to endow things with less mass than some other property does:

> S is a mass-property if and only if (1) S is a proto-mass-property, and (2) it is a law that for any x, and any proto-mass-properties P and Q, if x has property Q, and part of x has property P, then P endows things with less proto-mass than Q does.
>
> P endows things with less mass than Q does if and only if (1) P and Q are mass-properties, and (2) P endows things with less proto-mass than Q does.

Given these revised accounts, it will follow that the mass of a part of something cannot be greater than the mass of the whole. Moreover, I believe that any other necessary properties can be incorporated by further, appropriate revisions.

All the other required relations can be introduced in similar fashion, so it would seem that this alternative approach can be carried through. The crucial question, however, is whether the introduction of these second-order relations enables us to avoid the difficulty to which Field's own treatment was exposed. Recall that the problem was that a relation such as heavier than—thought of as holding either between space-time points, or physical objects—appears to be an internal relation, and therefore nothing over and above the non-relational properties of the relata. Is the situation different in the case of, say, the relation that obtains between two mass-properties when one endows objects with more mass than the other? I believe that it is. For while it is a necessary truth that the property of having a mass of ten kilograms endows an object with more mass than the property of having a mass of two kilograms, this merely reflects the descriptions that are being applied to the mass-properties in question. The properties in question are not even mass-properties, let alone specific mass-properties, by virtue simply of their intrinsic natures: what is crucial is the laws into which they enter. As a consequence, the second-order relations, which are defined in terms of those laws, are genuine external relations, and so may be constituents in the states of affairs that are truth-makers for statements expressing more complex laws, such as that of gravitation.

To briefly sum up this section. I have argued, first, that more complex scientific laws are best viewed as second-order laws, and second, that the most promising way of getting clear about the universals that enter into the relevant truth-makers is by reformulating the theories in question along the lines advocated by Hartry Field. His specific approach does, however, appear metaphysically problematic in a certain respect, and I have tried to show how that difficulty can be avoided by means of an account that involves second-order relations rather than first-order ones.

— 3 —

Difficulties and Objections

In this section I shall consider five objections to the approach to laws advanced in Chapter 2. The first is that the account offered of the truth conditions of nomological statements is in some sense *ad hoc* and unilluminating. The second questions whether the idea of contingent relations among universals is really intelligible. The third is concerned with the question of whether the present account implies that there can be logical connections between distinct states of affairs. The fourth is that the analysis commits one to a very strong version of realism with respect to universals. The final objection is that the account offered places an unjustifiable restriction upon the class of nomological statements.

3.1.1 *Is the Analysis Really Illuminating?*

The basic thrust of the first objection is this: there is a serious problem about the truth conditions of statements of laws. The solution proposed is that there are relations—referred to as nomological—that hold among universals, and that it is states of affairs consisting of universals standing in such relations that function as truth-makers for nomological statements. How does this solution differ from simply saying that there are special facts— call them nomological—which are the facts that make nomological statements true? How is the one account any more illuminating than the other?

The answer is two-fold. First, to speak simply of nomological facts does nothing to *locate* those facts, that is, to specify the individuals that are the constituents of the facts in question. In contrast, the view advanced here does locate the relevant facts: they are facts about universals, rather than facts about particulars. Moreover, support was offered for this contention, namely, that otherwise no satisfactory analysis of the truth conditions of statements of basic laws that lack positive instances is forthcoming.

Second, the relevant facts were not merely located, but *specified*, since not only the individuals involved, but their relevant attributes, were described. It is true that the attributes had to be specified theoretically, and hence in a sense indirectly. But this is also the case when one is dealing with theoretical terms attributing properties to particulars. The Lewis-type account of the meaning of theoretical terms that was appealed to is as applicable to terms that refer to relations among universals—including nomological relations—as to terms that refer to properties of, and relations among, particulars.

Still, the feeling that there is something unilluminating about the account may persist, and perhaps for two reasons. In the first place, one may very well feel that even though nomological relations can be specified in the indirect fashion employed above, it would be nice if something could be said about the *intrinsic* nature of such relations. Can anything be done in this regard? This is a question that I shall turn to in section 3.2, where I shall attempt to sketch an admittedly somewhat speculative metaphysical theory which seems to me promising, and which, if sound, will certainly throw light upon the intrinsic nature of nomological relations.

In the second place, there is the epistemological problem. How does one determine, after all, that there is, in any given case, a nomological relation holding among universals? In what sense have truth conditions for nomological statements really been supplied if it remains a mystery how one answers the epistemological question?

I think this is a legitimate issue, even though I do not accept the verificationist claim that a statement has factual meaning only if it is in principle verifiable. In section 3.3 I shall therefore turn to the epistemological problem raised by laws, and I shall attempt to show that, given the present account of the truth conditions of nomological statements, it is possible to have evidence that makes it likely that a given generalization is nomologically true.

3.1.2 *Can There Be Contingent Relations Among Universals?*

A second problem that arises for the present account concerns the fact that the truth-makers for law-statements are to be *contingent* relations among universals. The obstacle here is that many philosophers believe that there are grounds for grave doubts about the possibility of relations that are both contingent and, at the same time, genuinely relations simply among universals.

The picture seen by many philosophers who accept the existence of universals is, I think, as follows. On the one hand, there are what might loosely be spoken of as relations among universals, but which are really relations among the particulars that have the relevant properties. An example of this is the so-called relation of co-instantiation. Suppose that redness and roundness are co-instantiated. If this is so, it is certainly a contingent fact. The problem is that what makes this a fact is not some relation between the universals, redness and roundness. It is, rather, merely the fact that there are no particulars that have one of those properties, while lacking the other.

On the other hand, there do appear to be relations that are genuinely relations simply among universals, and that are not reducible to facts about particulars. Thus, when one says that a specific shade of red resembles a specific shade of orange more than it resembles a certain shade of blue, the truth of this claim does not seem to depend on any facts about particulars. So why may not a relationship such as resemblance provide us with a reason for admitting relations of the sort that are needed to provide truth-makers for statements of laws?

There are two problems here, both of which arise from the fact that resemblance is an *internal* relation. The first was mentioned in the discussion of more complex laws in section 2.3, namely, that there is reason for holding that internal relations are nothing over and above the related properties, with the result that it would seem that internal relations cannot be constituents of truth-makers. The second problem, and the one that I think should be stressed here, is that if two universals stand, so to speak, in the relation of resemblance, then it is logically necessary that they do so. In contrast, the relations among universals that are the truth-makers for statements of laws must be contingent, and hence external relations, if one is to avoid adopting the view that law statements express logically necessary truths.

In short, the situation is this. Those who hold that there are universals are generally happy with admitting relations (or apparent relations) among universals of the above two sorts. Many, however, seem to have serious doubts concerning the possibility of *external* relations which are genuine relations holding *simply* among universals. And it is precisely relations of this latter sort that are needed on the present view of laws.

It would be nice, then, if one could point to other, genuine, external relations among universals. For then one could break this stage of the argument down into two steps. First, one could argue that there can be, in general, external relations among universals. Second, one could then attempt to show that there can also exist a special subclass of such relations—namely, nomological ones.

I rather suspect, however, that any external relations among universals which are genuine relations among universals and nothing else, must be nomological relations. My reason for thinking that this is so is that I am inclined to accept a variation of the principle which Armstrong refers to as *the principle of descent to first-order particulars*: '. . . for any higher-order property or relation of order N, it is not simply instantiated by universals of order N-1, but is reflected in the particulars falling under these latter universals, and so on down until first-order particulars are reached'.[1]

Thus formulated, the principle incorporates the requirement that every universal be instantiated. I should prefer a conditionalized version of the principle, which states only what must be the case *if* a given universal is instantiated.

The line of thought that inclines me to the view that all genuine, external relations among universals will be nomological is then as follows: if the above principle of descent to first-order particulars, or a conditionalized version of it, is correct, then whenever one has a genuine relation among universals, there must be some corresponding fact that holds of each of the relevant particulars. Since it is a fact about every particular of the relevant sort, there must be a certain *generalization* about particulars which is true, by virtue of the higher-order relation among universals. If the higher-order relation is an internal and hence necessary one, then the generalization about particulars will be logically necessary—assuming that the principle of descent to first-order particulars is a logical truth. On the other hand, if the relation among universals is

[1] Armstrong, *Universals and Scientific Realism*, ii, p. 170.

an external and hence contingent one, then it seems likely that the generalization about particulars will be only contingently true. It would then seem reasonable to regard it as expressing a law of nature.

It seems likely, therefore, that the class of genuine, external relations among universals coincides with the class of nomological relations. So it is probably not possible to smooth the way for the view that the truth-makers for nomological statements are contingent relations among universals by pointing to other, unproblematic examples of contingent relations among universals, and then arguing that there exists a special subclass of such relations—namely, the nomological ones. This is unfortunate, since in the absence of clear-cut and uncontroversial examples of contingent relations among universals, doubts may linger concerning the ultimate coherence of the very idea of contingent relations among universals. Perhaps this idea, like that of a necessarily existent being, is really incoherent?

I should not deny that there are grounds for concern here. None the less, my feeling is that it is reasonable to assume that a concept is coherent unless there are definite reasons for thinking otherwise, and as there do not seem to be any in the case of the idea of contingent relations among universals, I believe that one is justified in making use of that concept—especially if it can be shown, as I shall attempt to do, that the resulting theory has a number of very significant virtues.

In addition, however, I do believe that it is possible to set out a theory of universals that, conjoined with an appropriate account of the intrinsic nature of nomological relations, allows one to explain how it is that universals can stand in contingent relations. This approach will be sketched in section 3.2.

3.1.3 *Logical Connections Between Distinct States of Affairs?*

A third objection, and one which has been stressed by David Lewis,[2] is this: how, exactly, are we to think of the relationship which purportedly obtains, on the present account of laws, between statements asserting that universals stand in certain nomological relations, and corresponding generalizations about

[2] In correspondence with David Armstrong. For Armstrong's discussion of this problem, see *What Is a Law of Nature?*, chap. 6, sect. III.

the properties and/or relations of first-order particulars? The relation is to be one of logical entailment. But is it a formal relation, or does one have to postulate *de re* connections between distinct states of affairs?

It might be thought that the answer to this is clear-cut. For given that laws are, on the present view, second-order states of affairs involving relations among universals which logically necessitate corresponding statements about first-order particulars, and given that first-order and second-order states of affairs are surely distinct, does not the above account of laws commit one to holding that there can be logical connections between distinct states of affairs?

I believe, however, that this line of thought is mistaken. For suppose that properties P and Q stand, say, in the relation of nomic necessitation. What first-order state of affairs is logically necessitated by this second-order state of affairs? The force of the question is perhaps most evident when one is dealing with a law that has no instances. But let us set aside that controversial case, and imagine that there is some particular, a, that has properties P and Q. Then the relation between the properties P and Q does not logically necessitate the first-order state of affairs which consists of a's having properties P and Q: the law would have obtained if a had not had property P, or if a had never existed. Now if one accepted the existence of conditional states of affairs, there would be a ready answer. The relevant first-order states of affairs would be states such as that which obtains whenever it is the case that if a exists and has property P, it also has property Q. But surely there is good reason not to postulate such states of affairs. And if that is so, then it is hard to see what first-order *states of affairs*—in contrast to first-order *sentences*—are logically necessitated by P's standing in the relation of nomic necessitation to Q.

If one thinks in terms of states of affairs, then, one cannot think in terms of a relation of necessitation between second-order and first-order states of affairs. What one has, rather, is a relation of necessitation that obtains between, on the one hand, a complex state of affairs that involves both a first-order part and second-order part, and, on the other, a first-order state of affairs. Thus, for example, the complex state of affairs which involves a's having property P, together with P's standing in the relation of nomic necessitation to Q, logically necessitates a's having property Q.

But is this situation any more easily accepted than the idea that there can be logical connections between first-order and second-order states of affairs? The question is not, I think, an easy one. There are, however, at least two ways that one might attempt to approach the problem. The one involves considering analogous situations that do not involve nomological relations. Suppose, for example, that property P resembles property Q. Then it is plausible to say that the complex state of affairs which involves the second-order state of P's resembling Q, together with the first-order state of a's having property P and b's having property Q logically necessitates the first-order state of a's resembling b. And thus one might argue that since this situation is acceptable, there is no reason not to accept the corresponding situation in the case of nomological states of affairs.

It might be questioned, however, whether these situations are really analogous. In particular, it might be argued that to say that a resembles b is just to say that there are monadic universals P and Q such that a has P, b has Q, and the universal P resembles the universal Q—which would imply that the entailment in the resemblance case, in contrast to the nomological case, is a formal one.

It is not clear, however, that this rejoinder is satisfactory. For is it really the case that to say that one particular resembles another is to say that they involve resembling universals? One reason for doubting that this is so is the fact that many philosophers have been attracted to the position known as resemblance nominalism—a position that is rendered immediately incoherent by the above account of what it means to say that two particulars resemble each other. It would seem preferable to search for an account of resemblance that is neutral with respect to the nominalism/realism debate. If such an account is possible, it would seem likely that the difficulty discussed in this section can be viewed as simply a case of a more general and difficult issue concerning logical interrelations among first-order and second-order states of affairs.

The other approach, and the one that I consider more promising, involves trying to construct a theory of universals that will make it clear how the problematic relation can be one of logical necessitation. This alternative tack will be considered in section 3.2.

3.1.4 *The Question of Transcendent Universals*

A fourth important objection to the view that laws are relations among universals concerns a rather serious metaphysical implication that this view appears to have. Let us say that a universal is transcendent if it is such as could exist even without having any instances, and let us construe Platonic realism in a minimal sense in which it involves only the claim that there are some transcendent universals. Then there is an argument which seems to show that even if the present account of laws does not entail that Platonic realism is true, it does entail that if the world had been slightly different, Platonic realism *would* have been true, and hence, that Platonic realism is at least a coherent position. As this conclusion would surely be unpalatable to many philosophers, we seem to have a possibly serious difficulty for the view that laws are relations among universals.

But why should the present view of laws entail anything about the existence of transcendent universals? Recall the discussion in section 2.1, concerning the possibility of underived laws that do not have instances of all possible types. Two cases were discussed: that of the fundamental particles, and that of the psychophysical laws. The present argument rests upon the second of these.

The relevant conclusions in the case of psychophysical laws were as follows. First, if reductive materialism is in fact false, and mental states sometimes involve emergent properties, then there is a possible world that differs from the actual world only with regard to the initial distribution of matter and energy, and in which there are basic, psychophysical laws that are expressed by vacuously true generalizations. Secondly, that if it is at least logically possible for reductive materialism to be false, then it is also possible for there to be worlds, which may be very different from the actual one, that contain basic, psychophysical laws that are expressed by vacuously true generalizations.

The consequence concerning transcendent universals now follows very quickly. Assume, for simplicity, not only that reductive materialism is false, but that the representative theory of perception is true, and that there is, in particular, a basic psychophysical law connecting certain sorts of brain state, as causes, to sensory experiences that have an emergent property of redness (of a specific shade)—where this property is not, of course, to be identified with that property of physical objects by virtue of which certain objects normally appears red. Given these assumptions, together with the first of the two conclusions mentioned above, one can say that in the world where life never evolves because the earth is too close to the sun, it will still be a basic psychophysical law that whenever organisms are in a certain complex brain state, they have experiences with a certain specific quality of redness, even though such a world contains, by hypothesis, no sentient organisms, and, *a fortiori*, no experiences of the red variety. But if laws are relations among universals, then it cannot be a basic law that whenever organisms are in a certain sort of brain state, they have experiences with a certain quality of redness, unless the relevant universals exist, including that property (of experiences) of having a certain quality of redness. Therefore, if laws are relations among universals, it is possible to describe a world that would contain uninstantiated, and hence transcendent, universals, since it would be a world which contained, for example, that property (of experiences) of having a certain quality of redness, even though there would be nothing in that world which possessed the relevant quality.

Thus put, the argument rests upon very controversial assumptions. But one can formulate the argument in terms of much weaker ones—such as the assumption that it is *logically possible* that reductive materialism is false, or the assumption that it is *logically possible* for there to be emergent properties. Given either of these claims, one can show that once it is admitted that there can be certain sorts of basic laws that fail to have instances of all possible sorts, it follows from the present account of laws that it is possible for there to be a world that contains transcendent universals. So Platonic realism, minimally construed, is at least a coherent position.

This is not a conclusion that will commend itself to most philosophers who reject Platonic realism, since the arguments directed against Platonic realism, if sound, are such as show that it is *necessarily* false. We need to consider, then, whether the conclusion can be avoided. One possibility is simply to deny that there can be emergent properties. This seems like a difficult doctrine indeed. If one admits both causal relations, and simple properties, one will surely have to admit the possibility that some complex property P might be such that instances of it are both causally necessary and causally sufficient for instances of some simple property Q—in which case Q would be an emergent property. And even if one rejects simple properties, it seems that one must still admit the possibility of emergent properties. For suppose that Q is some (possibly infinite) conjunctive property: Q_1 & Q_2 & ... & Q_i & ... It is hard to see how any contradiction could follow from instances of P being not merely causally sufficient and causally necessary for instances of Q, but also causally necessary for instances of every property Q_i which enters into Q. And if this were the case, Q would be an emergent property.

It does not seem very plausible, then, to respond to the argument sketched above—in support of the claim that Platonic realism is a coherent position—by denying the possibility of emergent properties. How, then, is one to respond to the argument? The response which Armstrong favours involves rejecting Platonic realism as a position that is *necessarily* false, and insisting upon what he refers to as the Principle of Instantiation: 'For each n-adic universal, U, there exist at least n particulars such that they are U.'[3]

One then draws the appropriate conclusion, to wit, that in the universe described above, where life never evolves, there simply will not be any psychophysical laws, since laws are relations among universals, and the relevant universals will not exist in the world in question.

It is possible, therefore, to accept the view that laws are relations among universals without committing oneself to the view that Platonic realism is a coherent position. But in order to do this, one has to reject a claim concerning the possibility of basic laws lacking relevant instances which many philosophers find appealing. So either way, it appears that the view that laws are relations among universals is in trouble:

(1) If laws are relations among universals, then *either* (i) there are at least some generalizations that cannot express basic laws unless they have instances of all possible types, *or* (ii) it is logically possible for there to be uninstantiated universals, and Platonic realism is therefore at least a coherent position.

(2) Any generalization that can express a basic law can do so

---

[3] Armstrong, *Universals and Scientific Realism*, i, p. 137.

even if the law fails to have instances of all possible types.

(3) Platonic realism is not a coherent position.

Therefore:

(4) Laws cannot be relations among universals.

This line of argument has been defended by Hugh Mellor.[4] Is there any satisfactory answer to it? We have already seen that there does not seem to be any plausible way of rejecting (1). Any successful challenge to the argument, therefore, must be directed against either (2) or (3). So let us consider the support that Mellor offers for these two premises.

In his discussion of what he refers to as vacuous laws, Mellor points to Newton's first law of motion, and to 'a multitude of nonanalytic laws quantifying over determinate values of continuously variable determinables'.[5] Mellor takes it that these cases show the importance, in science, of laws which may lack certain sorts of instances, since, on the one hand, it seems likely that Newton's first law of motion is itself vacuously true, and on the other, each of the laws which quantifies over determinate values of continuously variable determinables will entail, for each of those infinitely many values that are never realized, a vacuously true law.

The problem with such examples, however, is that they do not provide support for (2), since they are all cases of derived laws, whereas (2) is a thesis concerning basic laws. And in general, I suspect that this will probably be true of all actual laws—that is, that any laws that lack instances of certain sorts will, as a matter of fact, be derivable from laws that have instances of the relevant sorts. If this is right, then (2) cannot be supported by any appeal, such as Mellor makes, to actual scientific laws.

It needs to be noted, however, that this criticism of Mellor does depend, and crucially, upon the view that is taken of nomological generalizations involving quantification over determinate values of continuously variable determinables. Consider, for example, Newton's second law of motion. We saw, at the beginning of section 2.3, that there are two different views that one can take. First, one might hold that the generalization in question, rather

[4] Mellor, 'Necessities and Universals in Natural Laws', *Science, Belief, and Behaviour*, Cambridge, 1980, pp. 105-25.

[5] Ibid., p. 110.

than expressing a single law, expresses a class of related laws—one corresponding, perhaps, to each pair of possible values of mass and force. On this view, it will not be the case that there is a single state of affairs that is the truth-maker for a generalization expressing Newton's second law of motion. Rather, each of the individual laws will be a distinct state of affairs.

If this view were correct, it would not be true that Newton's first law of motion is a derived law. It would instead be a collection of basic laws—one for each possible mass—and there would have to be a nomological state of affairs corresponding to each of those basic laws. The support that Mellor offers for (2) would then be satisfactory.

The situation is very different if the correct view is the second one mentioned at the beginning of section 2.3. For on that view, laws such as Newton's second law of motion are indeed single laws, but they are second-order laws, and this has the consequence that the nomological states of affairs with which such laws are to be identified do not involve relations among universals some of which are uninstantiated. The laws to which Mellor appeals—such as Newton's first law of motion—will then have the status of derived laws, and the support which Mellor offers for (2) crumbles.

It was argued above, however, that the second view is the correct one to take. If that is right, then the conclusion seems to be that (2) cannot be supported by appealing to actual laws. One has to appeal to intuitions concerning *hypothetical* cases, of the sort discussed in section 2.1. This admittedly weakens the case in support of (2), but I believe that it is still very difficult to resist.

Let us consider, then, whether one can respond by rejecting (3), rather than (2). What support does Mellor offer for (3)—that is, for the claim that Platonic realism is incoherent? The answer is that he appeals to the sort of view of particulars and universals which Ramsey advanced in his paper, 'Universals'.[6] According to this view it is *facts*—or as some would prefer to say, states of affairs—which are the ontologically basic stuff of reality, out of which both particulars and universals are abstractions. This means that both particulars and universals are in a sense 'incomplete' objects.[7]

---

[6] Ramsey, 'Universals', *The Foundations of Mathematics*, pp. 112–34.
[7] Ibid., p. 121. Compare the view advanced by Armstrong in *Universals and Scientific Realism*, i, pp. 110 ff.

Mellor claims that it follows from this view that there cannot be
any uninstantiated universals.[8] This contention seems mistaken,
however, for the following reason. The view that reality is made
up of facts, or states of affairs, is itself completely neutral with
respect to the existence of higher-order universals. This means
that there will be at least two different types of states of affairs:
first-order states of affairs, consisting of particulars having
properties, and standing in relations, and second-order states of
affairs, consisting of first-order properties and relations having
second-order properties and standing in second-order relations.
Then, when one appeals to the principle that particulars and
universals are only abstractions from states of affairs, one will not
be able to derive the conclusion that every universal must be
instantiated, since there may be universals that enter into second-
order states of affairs without entering into first-order ones.

What is crucial here, however, is not Mellor's inference, but the
underlying metaphysics upon which it rests—namely, the idea that
states of affairs are ontologically basic, *and* that both particulars
and universals are abstractions out of states of affairs. Over against
this metaphysics, I wish to set another, the central theses of which
are, first, that reality must be determinate, that is, that whatever
exists must have an intrinsic nature, and second, that anything
with an intrinsic nature is capable of independent existence.

According to this alternative conception, states of affairs may be
of three sorts. First, those that consist of the existence of an
individual possessing an intrinsic nature. Second, those that
consist of the possession of properties by such individuals. Third,
those that involve relations among two or more such determinate
individuals.

It follows from this that bare particulars cannot exist, but bare—
i.e. uninstantiated—universals can. The former are precluded
because only what has a determiniate, intrinsic nature can exist,
other than as an abstraction. Universals on their own, however, do
have an intrinsic nature, and there is therefore no bar to their
enjoying independent existence.

This latter metaphysics is, I believe, superior to Mellor's, for it
enables us to explain a certain crucial intuition that cannot be
accounted for given his view. For if one fails to assign a central role

to the idea of having a determinate nature, then, given that some states of affairs involve relations between individuals, there would seem to be no reason why there could not be individuals that stood in various relations, but which had no properties at all. But this, I suggest, is not really intelligible. Individuals cannot exist, and *a fortiori* cannot stand in relations, unless they possess properties by virtue of which they have a determinate, intrinsic nature.

To sum up: the thrust of this fourth objection to the view that laws are relations among universals is that acceptance of this view forces one to hold either that some generalizations cannot express basic laws unless they have instances of all relevant sorts, or that Platonic realism is at least a coherent position. Some philosophers, such as Mellor, feel that both alternatives are unacceptable, and therefore that the view that laws are relations among universals must be abandoned. I have agreed with Mellor regarding the unacceptability of the first alternative. But on the other hand, I have tried to show that the sort of argument which Mellor offers against the second alternative is unconvincing, and I have urged that one should in fact respond to Mellor's argument by accepting that alternative.

In evaluating this choice, it is crucial to keep in mind that the expression 'Platonic realism' is being used here in a *much* weaker sense than normally, since it includes any view that admits the possibility of uninstantiated universals. Acceptance of Platonic realism in this minimal sense may be combined with very different grades of Platonic involvement. It may be helpful to consider briefly three main possibilities.

1. *Factual Platonic realism*: Even though it need not be the case that every universal is instantiated, it is still a *contingent* matter which universals exist in a given world: universals may enjoy independent existence, but not necessary existence.

2. *Moderate Platonic realism*: The universals that exist in any given world are those that exist according to the previous position, *plus* any universals that can be constructed out of those universals in certain ways, such as by conjunction.

3. *Strong Platonic realism*: No universal is such that its existence is in any way dependent upon any contingent state of affairs.

There is a very sharp contrast between the final position and the other two. According to Strong Platonic realism, universals are necessarily existent entities, whereas according to the Factual and Moderate versions of Platonic realism, what universals exist is a contingent matter. This difference is important in the present context; for given that the commitment to necessarily existent entities is, for many philosophers, one of the most unappealing features of Strong Platonic realism, it is essential to notice that the expression 'Platonic realism' is here being used to cover positions that are free of that commitment.

If one adopts the view that laws are relations among universals, and also holds that any possible basic law could obtain even if it failed to have instances of relevant sorts, one has to accept some sort of Platonic realism. But which version? I believe that one should opt for Factual Platonic realism. First, there is the principle of parsimony known as Ockham's Razor which asserts that one should not postulate more entities than are necessary. Secondly, as we shall see in section 3.2, a certain version of Factual Platonic realism appears to illuminate and reinforce, in certain ways, the general theory of laws that has been set out above.

3.1.5 *Nomological Statements and Essential Reference to Individuals*

The final objection focuses upon a feature that the above account shares with most other accounts of laws, namely, the view that generalizations that express laws should be free of any essential reference to specific individuals. In many other accounts of laws this feature derives from a restriction that is explicitly incorporated into the account. In the present case, however, this feature flows from the central idea that laws are relations among universals, together with the view that properties, and relations, in the strict sense, do not contain particulars as constituents.

The thrust of the fifth objection is that there are statements that it would be natural, in some possible worlds, to view as expressing laws, but which violate the condition that such statements be free of any essential reference to specific individuals. Suppose, for example, the world were as follows: All the fruit in Smith's garden at any time are apples. When one attempts to take an orange into the garden, it turns into an elephant. Bananas so treated become apples as they cross the boundary, while pears are resisted by a force that cannot be overcome. Cherry trees planted in the garden bear apples, or they bear nothing at all. If all these things were true, there would be a very strong case for its being a law that all the fruit in Smith's garden are apples. And this case would be in no way undermined if it were found that no other gardens, however similar to Smith's in all other respects, exhibited behaviour of the sort just described.

Given the account of laws and nomological statements set out above, it cannot, in the world described, be a law, or even a nomological truth, that all the fruit in Smith's garden are apples. If relations among universals are the truth-makers for nomological statements, a statement that contains essential reference to a specific particular can be nomological only if entailed by a corresponding, universally quantified statement free of such reference. And since, by hypothesis, other gardens do not behave as Smith's does, such an entailment does not exist in the case in question.

What view, then, is one to take of the generalization about the fruit in Smith's garden, in the world envisaged? One approach is to say that although it cannot be a law in that world, that all the fruit in Smith's garden are apples, it can be the case that there is some property P such that Smith's garden has property P, and it is a law that all the fruit in any garden with property P are apples. So that even though it is not a nomological truth that all the fruit in Smith's garden are apples, one can, in a loose sense, speak of it as 'derived' from a nomological statement.

This would certainly seem the most natural way of regarding the generalization about Smith's garden. The crucial question, though, is whether it would be reasonable to maintain this view in the face of any conceivable evidence. Suppose that careful investigation, over thousands of years, has not uncovered any difference in intrinsic properties between Smith's garden and other gardens, and that no experimental attempt to produce a garden that will behave as Smith's does has been successful. Would it still be reasonable to postulate a theoretical property P such that it is a law that all the fruit in gardens with property P are apples? This issue is by no means an easy one. On the one hand, it would seem that, given repeated failures to produce gardens that behave as Smith's garden does, one might well be justified in concluding that if there is such a property P, it is one whose exemplification outside Smith's garden is nomologically impossible. But to draw this conclusion would not only be to postulate a strange sort of property: it would itself entail the existence of a law involving an individual in an essential way.

Some such line of thought, therefore, might incline one to conclude that it must be logically possible for there to be generalizations that express laws—in the strict sense—and which involve ineliminable reference to specific individuals. This conclusion would not, however, force one to abandon the general approach to the nature of laws set out above. For if there can be such nomological statements, it would seem that only relatively minor modifications will be needed in the present account in order to accommodate them. The definition of a construction function would have to be changed so that particulars, and not merely universals, could be elements of the ordered n-tuples that it takes as arguments, and the definition of a nomological relation would have to be similarly altered, so that it could be a relation among both universals and particulars. These alterations would result in an analysis of the truth conditions of nomological statements that allowed for the possibility of ineliminable reference to specific individuals. They would do so, moreover, without opening the door to accidentally true generalizations: both condition (2^*), and condition (2^{**})—set out at the end of section 2.2—appear to be sufficiently strong to block such counterexamples.

Nevertheless, I am inclined to think that, in the end, one probably has to reject the idea that there can be laws that involve specific individuals in an essential way. My reason is that it seems to me that, if possible, one would very much like to have, not merely a formally satisfactory account of the meaning of nomological statements, but some theory that throws light upon the intrinsic nature of nomological relations. In the next section, we shall see that there is reason for being hopeful with regard to the prospects for such a theory, *if* laws can be viewed as relations simply among universals. There does not appear to be any way, however, in which the theory to be outlined might be extended to cover the case of laws into which particulars enter in an essential way. Smith's garden may tempt us to believe that such laws are possible, but this appears, sadly, to be a temptation that is best resisted.

3.2 TRANSCENDENT UNIVERSALS AND THE NATURE OF LAWS

In section 3.1.4, I considered the problem that is posed, for the view that laws are relations among universals, by the apparent possibility of basic laws that lack positive instances. The view advanced was that one should admit that possibility as a genuine one, and do so by adopting what I called Factual Platonic realism, according to which, though the existence of universals is a contingent matter, some universals may be transcendent, in the sense of being nowhere instantiated. What I want to argue in the present section is that this admission of transcendent universals, which may at first appear slightly unappealing, has in fact some significant advantages. First, it may make possible a more illuminating account of the nature of laws, and one which will, in particular, enable one to see how universals can stand in contingent relations. Secondly, it may provide an answer to the question, discussed in section 3.1.3, of whether the present account of laws commits one to holding that there can be logical relations between distinct states of affairs.

How is the account of laws to be advanced in this section related to that developed in Chapter 2? The answer is that the account to be set out in this section is a speculative interpretation of the general account offered in section 2.2, and, as such, it differs from it in two important respects. First, though I have indicated why I think that one should adopt a particular view of universals, the general account is in fact compatible with *any* position that admits the existence of universals. The account to be developed here, in contrast, rests upon Factual Platonic realism, and, indeed, upon a special version of that position. Second, in the general theory, nomological relations are characterized in an indirect fashion, as relations among universals that function in a certain way. For the essence of the approach is that nomological terms are treated as theoretical terms, and a Ramsey/Lewis approach to the meaning of theoretical terms is then applied to them. The speculative theory, on the other hand, provides a more direct characterization, according to which what laws of nature there are is capable of being unpacked simply in terms of what universals there are, together with part–whole relations between universals.

Let us consider, then, the metaphysics that is needed for the more speculative theory. The crucial idea is a distinction between dependent and independent universals. Let us say that a universal is a *dependent* universal in a given world if and only if it exists, in that world, only as a part of one or more other universals. It is an *independent* universal in a world if and only if it exists in that world other than as a part of other universals. What is required, then, is a version of realism that is compatible not only with the existence of universals, but also with *any* universal's turning out to be a dependent universal.

Some versions of realism preclude the existence of dependent universals. Some realists, for example, such as Reinhardt Grossman, reject the idea of complex universals. But if all universals are simple, none is a part of any other, and, *a fortiori*, there cannot be any that exist only as parts of other universals.

Standard versions of Platonism, according to which universals enjoy necessary existence, and the relations among universals are also necessary, would also seem to rule out the existence of dependent universals. For let U be any monadic universal. It would seem that there will be a possible world containing at least two things that have U in common, but no other properties. If universals are to explain exact resemblance of particulars, U has to be a universal which, in *that* world, does not exist only as a part of other universals. So it is not a dependent universal in that world. But if the same universals exist in all possible worlds, and if the relations among them are also necessary, then U cannot be a dependent universal in *any* world.

Given the distinction between dependent and independent universals, the metaphysical theory of the nature of laws that I have in mind may be explained as follows. Consider a world where, though some things have property Q without having property P, it is a law that everything with property P also has property Q. Assume, further, that not all universals are simple, and that, in particular, conjunctive universals exist whenever two properties are co-instantiated. Then Q may very well be an independent universal, and it is, in any case, not just a part of the conjunctive universal, P and Q. But what about P? Does it exist, in that world, only as a part of the conjunctive universal, P and Q? If one were an Aristotelian realist, it would seem that one's answer would have to be affirmative. For, on that view, universals exist only in the particulars that exemplify them, and it would seem, given this, that to say that one universal is part of another is just to say that every instantiation of the one is a part of some instantiation of the other. But what if one rejects Aristotelian realism in favour of Factual Platonic realism, as I have urged? Then, given the admission of transcendent universals, it would seem to be an open question whether universal *P* exists only as part of the conjunctive universal, *P* and *Q*. But what I want to suggest is that one adopt a specific version of Factual Platonic realism according to which there can be worlds where some universals exist only as parts of other universals, and where, moreover, it is the case that if it is a law that everything with property *P* has property *Q*, then the universal, *P*, exists only as part as the conjunctive universal, *P* and *Q*.

Given this very specific version of realism, one can explain what it is for there to be a law that everything with property *P* has property *Q* in a way that goes beyond the very general and indirect account offered above. For one can say that for there to be such a law is just for it to be the case that the conjunctive universal, *P* and *Q*, exists, as does the universal, *P*, but where the latter exists only as part of the former.

To what extent is this account dependent upon an acceptance of Factual Platonic Realism? We saw, above, that any version of Strong Platonic realism which holds both that universals are necessarily existent entities, and that relations among them are also necessary, precludes the existence of dependent universals. Such a theory therefore rules out the present account. If, however, one adopted a slightly weaker Platonism, according to which, even though the same universals exist in all possible worlds, the relations between them may vary from one world to another, then it would be possible for one universal to exist only as part of other universals in one world, while also existing on its own in some other world. The present account would thus be compatible with that sort of Strong Platonism.

In contrast, it seems clear that no version of Aristotelian realism is compatible with the above account of the nature of laws. For while an Aristotelian realist can allow for the existence of dependent universals, the class of dependent universals will turn out to be too broad in certain respects. Consider, for example, a world where it is just a grand accident that everything with property *P* also has property *Q*. An Aristotelian realist who accepts conjunctive universals will have to say that, in that world, the universal which is property P exists only as a part of the conjunctive universal P and Q, since, as we saw above, if transcendent universals are rejected, one universal's being part of another in a given world must be just a matter of every instantiation of the one being part of some instantiation of the other. To attempt to combine Aristotelian realism with the above account of laws would, therefore, lead to a collapse of the distinction between laws and merely accidental regularities.

How easily can the above theory be extended to laws of other sorts? In some cases, the extension is straightforward, while in other cases, substantial revision appears called for. Consider, first, what might be called disjunctive laws—such as a law that everything with property P has either property Q or property R. That would be a law in any world where the universal P existed, but only as part of two conjunctive universals, namely P and Q and P and R.

Consider, next, exclusion laws—such as the law that anything with property P lacks property Q. If there are no negative universals, then this exclusion law cannot be explained in terms of the universal P existing only as a part of the conjunctive universal P and not-Q. What one will need to do is to quantify over conjunctive universals, and say that the existence of that exclusion law is a matter of there being no conjunctive universal that has, as parts, both the universal P and the universal Q.

But if this account of exclusion laws is to work, there must be conjunctive universals that correspond, not merely to all universals that are somewhere co-instantiated, but to all nomologically possible combinations of properties. At this point, one may begin to feel that the world of universals is becoming somewhat overpopulated. This feeling is likely to be especially strong if one was initially drawn to the idea that only simple universals exist.

It is not, I think, easy to know how much weight should be assigned to this worry. But what needs to be emphasized here is that, even if one decided, in the end, that the ontological complexity that would be required to handle exclusion laws was too great to be acceptable, it would not follow that the sort of theory set out in this section is to be jettisoned. For there are two very different ways of viewing the theory. On the one hand, one might view it as giving an account of the meaning of nomological statements. Alternatively, it might be viewed as providing an *intrinsic* account of the nature of laws *in a given world*. The complexity introduced by exclusion laws, if unacceptable, would provide a reason for scrapping the theory as an account of the meaning of nomological statements. For when it is so viewed, it is the mere possibility of exclusion laws that is relevant. In contrast, if the theory is interpreted as an account of the intrinsic nature of laws in a given world, the complexity that would be introduced by exclusion laws is of no concern if the world in question contains no laws of that sort.

I believe that the theory offered in the present section should *not* be viewed as an analysis of the meaning of nomological statements: the latter is provided by the general account developed in Chapter 2. That general account does place some restrictions upon the nature of laws. It implies that laws cannot be mere regularities. It implies that laws involve the existence of universals, and of contingent relations among universals. But, on the other hand, no account of the *intrinsic* nature of nomological relations is provided by that general account: the characterization is entirely relational. The account sketched here, in contrast, does address itself to the question of what those relations are in themselves. Thus it says, for example, that the relation of nomic necessitation that holds between universals P and Q, when it is a law that everything with property P has property Q, is simply a matter of its being the case that the universal P exists only as a part of the conjunctive universal P and Q. But in saying this, it need not be viewed as expressing a further, conceptual claim. Rather, it may be viewed as advancing a metaphysical hypothesis concerning what the relation of nomic necessitation consists of in at least some worlds—namely, those that do not suffer from the complexity generated by laws of certain sorts, such as exclusion laws.

The sort of theory suggested in the present section has a bearing on almost all of the objections discussed earlier in this chapter. First, consider the issue raised in section 3.1.5—the case of Smith's garden. We saw that there appeared to be no serious barrier to modifying the general account of laws to accommodate the idea that there could be laws of nature that involved some individual in an essential way. But if one considers such an apparent possibility with an eye to what account is to be given concerning the intrinsic nature of nomological relations, a less tolerant attitude may be appropriate. For it is certainly not easy to see how anything along the lines of the theory sketched here could be modified to allow for such laws.

Second, there is the issue discussed in section 3.1.1, of whether the analysis proposed in Chapter 2 is really illuminating. That issue can now be viewed in a somewhat different light, since one might now think that the underlying dissatisfaction was not really with the analysis *qua* analysis, but with the fact that no account was being offered of the intrinsic nature of nomological relations among universals.

Third, there is the question of the possibility of contingent relations among universals, which was canvassed in section 3.1.2. That issue also appears in a different light, given the discussion in the present section. For contingent relations of a sort sufficient for the general theory will be possible, provided that the following three conditions can be met: (1) it is possible for a universal to be part of another universal; (2) it is possible for a universal to exist which does so only as part of another universal; (3) it is possible for there to be transcendent universals. This, of course, does not settle the issue, for the third of these conditions is by no means completely unproblematic, the discussion in section 3.1.4 notwithstanding. But one can at least say that the idea of contingent relations among universals appears to be no more problematic than the thesis that there can be uninstantiated universals.

Finally, there is the problem posed by David Lewis and discussed in section 3.1.3, of whether the present approach to laws commits one to an acceptance of logical relations between distinct states of affairs. The question was what account could be given of the entailment that holds, for example, between the complex state of affairs which consists of a's having property P, together with its being a law that everything with property P has property Q, and the state of affairs which consists of a's having property Q. If laws are simply certain regularities, there is no difficulty. But if laws are relations between universals, then we must have a logical relation between, on the one hand, the conjunction of a first-order state of affairs and a second-order one, and, on the other, another first-order state of affairs. How can this be the case?

The theory developed in this section provides a satisfying answer to this problem. If it is a law that everything with property P has property Q, then the universal which is property P exists, in that world, only as a part of the conjunctive universal *P* and *Q*. Therefore, the only way that property *P* can be instantiated in a given individual, in a world where that laws obtains, is if the conjunctive universal *P* and *Q* is instantiated in that individual.

3.3 THE EPISTEMOLOGICAL QUESTION

In section 3.1.1 I referred to, but then set aside, the claim that the analysis of nomological statements advanced above is unilluminating because it provides no account of how one determines whether specific universals stand in a given nomological relation. I shall now attempt to show that this objection is unsound, and that in fact one of the merits of the present view is that it does make possible an answer to the epistemological question.

Suppose, then, that a statement is nomological if there is an irreducible theoretical relation holding among certain universals which necessitates the statement's being true. If this is right, why should there be any special epistemological problem about the ground for accepting a given nomological statement? To assert that such a statement is true will be to advance a theory claiming that certain universals stand in some nomological relation. So it is natural to suppose that whatever account is to be offered in general of the grounds for accepting theories as true should also be available in the present case.

I should like, however, to make plausible in a more direct fashion the suggestion that the sorts of considerations that guide our choices among theories also provide adequate grounds for preferring some hypotheses about nomological statements to others. Consder, then, a familiar sort of example: John buys a pair of trousers, and is careful to put only silver coins in the right-hand pocket. This he does for several years, and then destroys the trousers. What is the status of the generalization: 'Every coin in the right-hand pocket of John's trousers is silver'? Even on the limited evidence described, one would not be justified in accepting it as nomological. For one of the central reasons for accepting a given theory is the fact that it in some sense provides the best explanation of certain observed states; and while the hypothesis (that there is a theoretical relation holding among the universals involved in the proposition that every coin in the right-hand pocket of John's trousers is silver, which necessitates that proposition's being true) does explain why there were only silver coins in the pocket, this explanation is by no means necessary. Another explanation is readily available—that John wanted to put only silver coins in that pocket, and carefully inspected every coin to ensure that it was silver before putting it in.

If the evidence is expanded in certain ways, the grounds for rejecting the hypothesis that the statement is nomological become even stronger. Suppose that one has made a number of tests on other trousers, ostensibly similar to John's, and found that the right-hand pockets accept copper coins as readily as silver ones. In the light of this additional evidence, there are two main hypotheses to be considered:

(H_1): It is nomologically possible for the right-hand pocket of any pair of trousers of type T to contain a non-silver coin;

(H_2): There is a pair of trousers of type T, namely John's, such that it is nomologically impossible for the right-hand pocket to contain a non-silver coin; however all other pairs of trousers of type T are such that it is nomologically possible for the right-hand pocket to contain a non-silver coin.

(H_1) and (H_2) are conflicting hypotheses. Each is compatible with all the evidence, but it is clear that (H_1) is to be preferred to (H_2). First, because (H_1) is simpler than (H_2), and second, because the generalization explained by (H_2) is one for which we already have an explanation.

Let us now try to get clearer, however, about the sort of evidence that provides the strongest support for the hypothesis that a given generalization is nomological. I think the best way of doing this is to consider a single generalization, and to ask, first, what a world would be like in which one would feel that the generalization was merely accidentally true, and then, what changes in the world might tempt one to say that the generalization was not accidental, but nomological.

I have already sketched a case of this sort. In our world, if all the fruit in Smith's garden are apples, it is only an accidentally true generalization. But if the world were different in certain ways, one might classify the generalization as nomological. If one never succeeded in getting pears into Smith's garden, if bananas changed into apples, and oranges into elephants, as they crossed the boundary, etc., one might well be tempted to view the generalization as expressing a law.

What we now need to do is to characterize the evidence that seems to make a crucial difference. What is it about the sort of events described that makes them significant? The answer, I suggest, is that they are events that determine which of conflicting generalizations are true. Imagine that one has just encountered Smith's garden. There are many generalizations that one accepts—generalizations that are supported by many positive instances, and for which no counterexamples are known, such as 'Pears thrown with sufficient force toward nearby gardens wind up inside them,' 'Bananas never disappear, nor change into other things such as apples,' 'Cherry trees bear only cherries as fruit.' One notices that there are many apples in the garden, and no other fruit, so the generalization that all the fruit in Smith's garden are apples is also supported by positive instances, and is without counterexamples. Suppose now that a banana is moving in the direction of Smith's garden. A partial conflict situation exists in that there are some events that will, if they occur, falsify the generalization that all the fruit in Smith's garden are apples, and other events that will falsify the generalization that bananas never change into other objects. Other events are possible of course, which will falsify neither generalization: the banana may simply stop moving as it reaches the boundary. However there may well be other generalizations, which one accepts, that make the situation one of inevitable conflict, so that whatever the outcome, at least one generalization will be falsified. Situations of inevitable conflict can arise even for two generalizations, if related in the proper way. Thus, given the generalizations that $(x)(Px \supset Rx)$ and that $(x)(Qx \supset \sim Rx)$, discovery of an object b such that both Pb and Qb would be a situation of inevitable conflict.

Many philosophers have felt that a generalization's surviving such situations of conflict, or potential falsification, provides strong evidence in support of the hypothesis that the generalization is nomological. The problem, however, is to *justify* this view. One of the merits of the account of the nature of laws offered here is that it provides such a justification.

The justification runs as follows: suppose that one's total

evidence contains a number of supporting instances of the generalization that $(x)(Px \supset Rx)$, and of the generalization that $(x)(Qx \supset \sim Rx)$, and no evidence against either. Even such meagre evidence may provide some support for the hypothesis that these generalizations are nomological, since the situation may be such that the only available explanation for the absence of counterexamples to the generalizations is that there are theoretical relations holding among universals which necessitate those generalizations. Suppose now that a conflict situation arises: an object b is discovered which is both P and Q. This new piece of evidence will reduce somewhat the likelihood of both hypotheses, since it shows that at least one of them must be false. Still, the total evidence now available surely lends some support to both hypotheses. Let us assume that it is possible to make at least a rough estimate of that support. Let m be the probability, given the available evidence, that the generalization that $(x)(Px \supset Rx)$ is nomological, and n the probability that the generalization that $(x)(Qx \supset \sim Rx)$ is nomological—where $(m + n)$ must be less than or equal to one. Suppose finally that b, which has properties P and Q, turns out to have property R, thus falsifying the second generalization. What we are now interested in is the effect this has upon the probability that the first generalization is nomological. This can be calculated by means of Bayes's Theorem:

Let S be: It is a nomological truth that $(x)(Px \supset Rx)$.
Let T be: It is a nomological truth that $(x)(Qx \supset \sim Rx)$.
Let H_1 be: S and not-T.
Let H_2 be: Not-S and T.
Let H_3 be: S and T.
Let H_4 be: Not-S and not-T.

Let E describe the total antecedent evidence, including the fact that Pb and Qb.
Let E^* be: E and Rb

Then Bayes's Theorem states:

Probability (H_1, given that E^* and E) =

$$\sum_{i=1}^{4} \frac{\text{Probability } (E^*, \text{ given that } H_1 \text{ and } E) \times \text{Probability } (H_1 \text{ and } E)}{[\text{Probability } (E^*, \text{ given that } H_i \text{ and } E) \times \text{Probability } (H_i \text{ and } E)]}$$

Taking the antecedent evidence as given, we can set the

probability of E equal to one. This implies that the probability of H_i and E will be equal to the probability of H_i given that E.

Probability (H_3, given that E)
   $=$ Probability (S and T, given that E)
   $= 0$, since E entails that either not-S or not-T.

Probability (H_1, given that E)
   $=$ Probability (S and not-T, given that E)
   $=$ Probability (S, given that E) $-$
       Probability (S and T, given that E)
   $= m - 0 = m$.

Similarly, probability (H_2, given that E) $= n$.

Probability (H_4, given that E)
   $= 1 - $ [Probability (H_1, given that E) $+$
       Probability (H_2, given that E) $+$
       Probability (H_3, given that E)]
   $= 1 - (m + n)$

Probability (E^*, given that H_i and E)
   $= 1$, if $i = 1$ or $i = 3$
   $= 0$, if $i = 2$
   $=$ some value k, if $i = 4$.

Bayes's Theorem then gives:

Probability (H_1, given that E^* and E)

$$= \frac{1 \times m}{(1 \times m) + (0 \times n) + (1 \times 0) + (k \times (1 - (m + n)))}$$

$$= \frac{m}{m + k(1 - m - n)}$$

In view of the fact that, given that E^*, it is not possible that T, this value is also the probability that S, that is, the likelihood that the generalization that (x) $(Px \supset Rx)$ is nomological. Let us consider some of the properties of this result. First, it is easily seen that, provided neither m nor n is equal to zero, the likelihood that the generalization is nomological will be greater after its survival of the conflict situation than it was before. For the value of

$$\frac{m}{m + k(1 - m - n)}$$

will be smallest when k is largest, and setting k equal to one gives the value

$$\frac{m}{m + 1 - m - n}$$

i.e.,

$$\frac{m}{1 - n}$$

which is greater than m if neither m nor n is equal to zero.

Second, the value of

$$\frac{m}{m + k(1 - m - n)}$$

increases as the value of k decreases, and this too is desirable. If the event that falsified the one generalization would have been very likely even if neither generalization had been nomological, one would not expect it to lend as much support to the surviving generalization as it would if it were an antecedently improbable event.

Thirdly, the value of

$$\frac{m}{m + k(1 - m - n)}$$

increases as n increases. This means that survival of conflict with a well-supported generalization results in a greater increase in the likelihood that a generalization is nomological than survival of conflict with a less well-supported generalization. This is also an intuitively desirable result.

Finally, it can be seen that the evidence provided by survival of conflict situations can quickly raise the likelihood that a generalization is nomological to quite high values. Suppose, for example, that $m = n$, and that $k = 0.5$. Then

$$\frac{m}{m + k(1 - m - n)}$$

will be equal to $2m$. This result agrees with the view that statements of laws, rather than being difficult or impossible to confirm, can acquire a high degree of confirmation on the basis of relatively few observations, provided that those observations are of the right sort.

But how is this justification related to the account advanced above as to the truth conditions of nomological statements? The answer is that there is a crucial assumption that seems reasonable if relations among universals are the truth-makers for nomological statements, but not if facts about particulars are the truth-makers. This is the assumption that m and n are not equal to zero. If one takes the view that it is facts about the particulars falling under a generalization that make it the case that it expresses a law, then, if one is dealing with a possibly infinite universe, it is hard to see how one can be justified in assigning any non-zero probability to a generalization, given evidence concerning only a finite number of instances. For surely there is some non-zero probability that any given particular will falsify the generalization, and this entails, given standard assumptions, that as the number of particulars increases without limit, the probability that the generalization will be true approaches indefinitely close to zero.[9]

By contrast, if laws are relations among universals, the truth-maker for a given nomological statement is an atomic fact, and it would seem justified, given standard principles of confirmation theory, to assign some non-zero probability to this fact's obtaining. So not only is there an answer to the epistemological question; it is one that appears to be available only given the approach to laws advocated here.

One objection to this account of the confirmability of laws needs to be considered. The thrust of the objection is that one is *not* justified in assigning non-zero initial probability to a certain nomological state of affairs' being the case. The argument is this: consider a typical nomological generalization, such as that expressing the Newtonian law of gravitation:

$$F = \frac{k(m_1 \times m_2)}{r^2}$$

There are infinitely many generalizations that are incompatible with this. Most of them are considerably more complex. However

---

[9] See, for example, Carnap's discussion of the problem of the confirmation of universally quantified statements in *The Logical Foundations of Probability*, 2nd edn., Chicago, 1962, pp. 570–1.

there is also an infinite number of incompatible generalizations that are equally simple: one need merely assign different values to the constant term k. So if one is justified in assigning some non-zero initial probability to its being the case that the Newtonian law of gravitation obtains, it would seem that one must be justified in assigning the *same* non-zero initial probability to each of the infinite number of equally simple, incompatible laws. And this leads to a contradiction, since it entails that the probability that at least one of these incompatible nomological situations obtains is greater than one.

This objection is sound, but it does not undermine the basic line of argument offered above. What the objection shows, at least in the case of quantitative laws, is that if probabilities are represented by real numbers—rather than by real numbers plus infinitesimals—then no non-zero number can justifiably be chosen as representing the initial probability that a certain generalization is a law. But this does not rule out the possibility of assigning non-zero probabilities given one or more positive instances falling under the generalization. Consider the case of the Newtonian law of gravitation, and the unlimited number of equally simple, incompatible nomological generalizations that result when different values are substituted for k. Given a single case of gravitational attraction, all but one of those generalizations will be falsified. So no problem arises if one assigns some non-zero probability to the remaining generalization's being nomologically true, given that it has at least one positive instance.

But is it certain that no problem arises? Milton Fisk has attempted to show, by means of an argument which appeals to Bayes's Theorem, that in the case where there are an infinite number of incompatible nomological hypotheses, 'data can add support to an hypothesis only if there is an initial chance of the hypothesis being true'.[10] If this claim were correct, my response to the above objection would have to be unsound. It would not be possible for a nomological hypothesis whose initial probability was zero to have some non-zero probability upon the evidence that it has at least one positive instance. Fisk's argument, however, turns upon the assumption that 'even when the hypotheses are infinite in number, probabilities are by definition either zero or finite'.[11] It is

---

[10] Fisk, 'Are There Necessary Connections in Nature?', *Philosophy of Science*, 37/3, pp. 385–404. See p. 390.   [11] Ibid.

easily shown that this assumption must be rejected. Imagine a world that contains an infinite number of marbles. Let H_i be the hypothesis that only the ith marble is red, and let H be an abbreviation for the infinite disjunctive statement: H_1 or H_2 or ... or H_i or ... Assume finally that the probability of H_i given that H is equal to the probability of H_j given that H, for all i and j, and represent this value by k. Since H_i and H_j cannot both be true if i is different from j, it follows from the laws of probability that

$$\text{Prob}(H, H) = \text{Prob}(H_1, H) + \text{Prob}(H_2, H) + \ldots$$
$$+ \text{Prob}(H_i, H) + \ldots$$
$$= k + \ldots + k + \ldots$$

If k is some real number greater than zero, this infinite sum will be greater than one. If k is zero, the infinite sum will also be zero. So in neither case will the infinite sum be equal to the probability of H given that H, which is one. It follows, therefore, that the probabilities of the H_i, given that H, cannot be adequately represented by real numbers, contrary to Fisk's assumption. If probabilities are to be assigned in such cases, infinitesimals are needed if contradictions are to be avoided. And when these are introduced, Fisk's argument can be seen to be fallacious. It appears, then, that the account of the truth conditions of nomological statements advanced here does provide the basis for a satisfactory explanation of the confirmability of laws.

3.4 ADVANTAGES OF A REALIST APPROACH

Some positive features of the present approach to laws have been mentioned at various places in the above discussion. It may be helpful in this concluding section to recall some of the more significant points, and to mention some others that have not been touched upon.

In the first place, then, the present approach to laws answers the challenge advanced by Roderick Chisholm: 'Can the relevant difference between law and non-law statements be described in familiar terminology without reference to counterfactuals, without use of modal terms such as "causal necessity", "necessary condition", "physical possibility", and the like, and without use of metaphysical terms such as "real connections between matters of fact"?'[12] The account offered does precisely this. There is no reference to counterfactuals. The notions of logical necessity and logical entailment are used, but no nomological modal terms are employed. Nor are there any metaphysical notions, unless a notion such as a contingent relation among universals is to be counted as metaphysical.

A second advantage of the account is that it contains no reference to possible worlds. What makes nomological statements true are not facts about dubious entities called possible worlds, but facts about the actual world. True, these facts are facts about universals, not about particulars, and they are theoretical facts, not observable ones. But neither of these things should worry one unless one is either a reductionist, at least with regard to higher-order universals, or a rather strict verificationist.

Third, the account provides a clear and straightforward answer to the question of the difference between nomological statements and accidentally true generalizations. A generalization is accidentally true by virtue of facts about particulars: it is nomologically true by virtue of a relation among universals.

Fourth, this view of the truth conditions of nomological statements explains the relation between counterfactuals and different types of generalizations. Suppose that it is a law that $(x)(Px \supset Qx)$. This will be so if the relevant universals stand in the appropriate relation. Let us now ask what would be the case if some object, b, which presently lacks property P, were to have that property. In particular, is it reasonable to assume that it would still be a law that $(x)(Px \supset Qx)$? It seems reasonable to reply that the supposition about the particular object, b, does not give one any reason for concluding that properties P and Q no longer stand in the relation of nomic necessitation. If this is right, then one can justifiably conjoin the supposition that b has property P with the proposition that the nomological relation in question holds between properties P and Q, from which it will follow that b has property Q. And this is why one is justified in asserting the counterfactual: If b were to have property P, it would also have property Q.

Suppose, by contrast, that it is only an accidentally true

---

[12] Chisholm, 'Law Statements and Counterfactual Inference', *Analysis*, 15, 1955, pp. 97–105. Reprinted in *Causation and Conditionals*, ed. Sosa, Oxford, 1975. See p. 149.

generalization that $(x)(Px \supset Qx)$. Then it is facts about particulars that make the generalization true. As a consequence, if one asks what would be the case if some particular b, which lacks property P, were to have P, the situation is very different. Now one is supposing an alteration in facts that may be relevant to the truth conditions of the generalization. So if object b lacks property Q, the appropriate conclusion may be that if b were to have property P, the generalization that $(x)(Px \supset Qx)$ would no longer be true, and thus that one would not be justified in conjoining that generalization with the supposition that b has property P in order to support the conclusion that b would, in those circumstances, have property Q. And this is why accidentally true generalizations, unlike laws, do not support the corresponding counterfactuals.

Fifth, this account of nomological statements allows for the possibility of even basic laws that lack positive instances—an outcome which, I suggest, accords well with our intuitions about what laws there be in cases such as the slightly altered version of our own world, where life never evolves, and in that of the universe with the two types of fundamental particle that never meet.

Sixth, it is a consequence of the account given that if S and T are logically equivalent sentences, they must express the same law, since there cannot be a nomological relation among universals that would make the one true without making the other true. I believe that this is a desirable consequence. However some philosophers have contended that logically equivalent sentences do not always express the same law. Nicholas Rescher, for example, in his book *Hypothetical Reasoning*, claims that the statement that it is a law that all Xs are Ys makes a different assertion from the statement that it is a law that all non-Ys are non-Xs, on the ground that the former asserts not only that all Xs are Ys but also that if Z (which isn't an X) were an X, then Z would be a Y; while the latter asserts not only that all non-Ys are non-Xs, but also that if Z (which isn't a non-Y) were a non-Y, then Z would be a non-X.[13] But it would seem that the answer to this is simply that the statement that it is a law that all Xs are Ys *also* entails that if Z (which isn't a non-Y) were a non-Y, then Z would be a non-X. So Rescher has not given us any reason for supposing that logically equivalent sentences can express different laws.

[13] Rescher, *Hypothetical Reasoning*, Amsterdam, 1964, p. 81.

The view that logically equivalent sentences can express different laws has also been advanced by Stalnaker and Thomason. Their defence of this contention rests, however, upon the idea that law-statements are equivalent to generalized subjunctive conditionals— a view that was shown to be unsound in section 2.1.2.

A seventh point is that given the above account of laws, it is easy to show that nomological statements have the logical properties one would normally attribute to them. Contraposition holds for statements of laws, and more generally, for nomologically true statements, by virtue of the fact that logically equivalent statements express the same law. Transitivity also holds: if it is a law (or nomological truth) that $(x)(Px \supset Qx)$, and also that $(x)(Qx \supset Rx)$, then it is also a law (or nomological truth) that $(x)(Px \supset Rx)$. Moreover, if it is a law (or nomological truth) that $(x)(Px \supset Qx)$, and also that $(x)(Px \supset Rx)$, then it is a law (or nomological truth) that $(x) (Px \supset (Qx \& Rx))$, and conversely. Also, if it is a law (or nomological truth) both that $(x)(Px \supset Rx)$ and that $(x)(Qx \supset Rx)$, then it is a law (or nomological truth) that $(x) ((Px \lor Qx) \supset Rx)$, and conversely. And in general, I think that law-statements and nomological statements can be shown, on the basis of the analysis proposed above, to have all the formal properties they are commonly thought to have.

An eighth point is that the account provides a straightforward explanation of the non-extensionality of nomological contexts. The reason that it can be a law that $(x)(Px \supset Rx)$, and yet not a law that $(x)(Qx \supset Rx)$, even if it is true that $(x)(Px \equiv Qx)$, is that, in view of the fact that the truth-makers for nomological statements are relations among universals, the referent of the predicate P in the sentence: It is a law that $(x)(Px \supset Rx)$, is—in the simplest case—the associated universal, rather than the set of particulars falling under the predicate. As a result, interchange of co-extensive predicates in nomological contexts may alter the referent of part of the sentence, and with it the truth of the whole.

Finally, various epistemological issues can be resolved given this account of the truth conditions of nomological statements. How is it possible to establish that a generalization expresses a law, rather than merely being accidentally true? The general answer is that if law-statements are true by virtue of theoretical relations among universals, then whatever account is to be given of the grounds for accepting theories as true will also be applicable to statements of laws. The latter will not pose any independent problems. Why is it that the results of a few carefully designed experiments can apparently provide very strong support for a nomological statement? The answer is that if the truth-makers for law-statements are relations among universals, rather than facts about particulars, the assignment of non-zero initial probability to a nomological statement's being true ceases to be unreasonable, and one can then employ standard theorems of probability theory, such as Bayes's Theorem, to show how a few observations of the right sort will result in a probability assignment that quickly takes on quite high values.

It would seem, then, that the view that laws are contingent, irreducible, theoretical relations among universals has much to recommend it. For it provides, first, a non-circular, extensional account of the truth conditions of nomological statements; second, an explanation of the formal properties of such statements; and third, a solution to epistemological problems concerning the confirmability of hypotheses concerning laws.

— 4 —

Probabilistic Laws

This chapter is concerned with the question of the nature of probabilistic laws. In the first section we shall see that the problem of truth-makers for nomological statements appears to be even more serious in the case of statistical laws than in the case of non-statistical laws; and also, that there appears to be little prospect for a satisfactory solution to this problem unless contingent, irreducible relations among universals are taken as the truth-makers for such statements. I shall therefore attempt to show how the approach to non-probabilistic laws set out and defended in Chapters 2 and 3 can be extended, in a natural and relatively straightforward way, to probabilistic laws.

4.1 THE TRUTH-MAKERS FOR STATEMENTS EXPRESSING PROBABILISTIC LAWS

It was argued in section 2.2 that non-probabilistic laws may lack certain sorts of instances, and that, consequently, the truth-makers for law statements must be relations among universals. The first point here, then, is that precisely the same argument applies to statistical laws.

The second point is related to an example mentioned in section 2.1, namely, the case of a world where it is a statistical law that the probability that something with property P has property Q is 0.999999999, but where there are very few things with property P, and it happens that all of them have property Q. This example might have been used to support the claim that the truth-makers for statements expressing non-statistical laws are relations among universals. For consider a world in which everything with property P also has property Q, and where the connection is not merely accidental. Perhaps it is a law that everything with property P has property Q. But perhaps the relevant law is rather that the probability that something with property P has property Q is, say, 0.999999999. If the former is in fact the case, rather than the latter, what is it that makes this so? What is the relevant truth-maker?

I believe that this question gives rise to a strong argument in support of the claim that it is relations among universals which are the relevant truth-makers. Given the bias that it is rather natural to have in favour of non-statistical laws, however, the argument, as formulated above, might not have seemed to carry much weight. The case is, I think, rather different when the choice is not, as above, between a statistical law and a non-statistical law, but among competing probabilistic laws. Imagine a world, then, where there are only four things with property P, and three of them have property Q. Perhaps it is a law that the probability that something with property P has property Q is 0.75. But one certainly does not want to hold that the fact that three out of four things with property P have property Q precludes other possibilities, such as its being a law that the probability that something with property P has property Q is 0.8. For that would imply, for example, that whether or not it can be a law that the probability that something with property P has property Q is 0.75 depends upon how many things with property P the world happens to contain. If it contains four such things, it may be a law. If it contains five, it cannot be. It would also imply that it could not be a law, in any world containing only a finite number of things with property P, that the probability that something with property P has property Q is some irrational number, such as $\pi/4$. Such consequences seem unacceptable. The conclusion, therefore, is that relative frequencies within *actual* events cannot serve as the truth-makers for statements expressing probabilistic laws.

What, then, are the truth-makers for statements expressing statistical laws? What makes it a law, for example, that the probability that something with property P has property Q is 0.75, rather than $\pi/4$?

Two popular views need to be examined. The first appeals to the notion of the limiting value of some suitably defined relative frequency. Let S be some infinite sequence of individuals possessing property P, where the sequence is randomly generated, so that, for example, the selection of elements for sequence S does not depend upon whether an individual has property Q. Let S_i be the set consisting of the first i elements, and let $N(i)$ be the number of elements in S_i that have property Q. If the ratio $N(i)/i$ approaches some limit k as i tends to infinity, then it is said, on this first approach, that the probability that something with property P has property Q is equal to k. It is facts about randomly generated infinite sequences which determine, therefore, what probabilistic laws there are.

There are various objections to this approach. The most important, in my opinion, is that it fails to take seriously the idea that statements cannot be true unless they have truth-makers, where truth-makers must be states of affairs consisting of properties of, and relations among, *actually existing things*. For when it is said that probability statements are to be interpreted in terms of the limits of relative frequencies in appropriately defined, infinite sequences of events, this is not to be taken as implying that there exists a probability that objects with property P have property Q only if there is an infinite sequence of *actual* objects possessing property P. To say that the probability that objects with property P have property Q is equal to k is being construed, rather, as saying only that if there *were* certain appropriate infinite sequences, then the limit of the relative frequency within those sequences *would* be equal to k. The semantical interpretation of probability statements which is offered by the advocate of the limiting relative frequency conception is in terms, then, of 'possible' events or objects. As a consequence, this approach fails to supply truth-makers for probabilistic statements.

A second problem is this: even where there is an appropriately random, infinite sequence of actual events, it cannot be correct to *define* the probability in terms of the limit of the relative frequencies. For suppose that one has a fair coin, and that it is flipped, in some sort of unbiased fashion, an infinite number of times. Does it *follow* from this that the limit of the relative frequency of heads to total tosses will be 0.5? The answer, as a number of writers have emphasized, is that it does not.[1] What is true is that given positive numbers d and e, no matter how small, one can find some number N such that for any initial segment of

[1] See, for example, p. 225 of Railton's article 'A Deductive–Nomological Model of Probabilistic Explanation', *Philosophy of Science*, 45, 1978, pp. 206–26; p. 270 of Lewis's 'A Subjectivist's Guide to Objective Chance', *Studies in Inductive Logic and Probability*, ii, ed. Jeffrey, Berkeley and Los Angeles, 1980, pp. 263–93; and, for a more detailed discussion, sect. IA6 of Skyrms's *Causal Necessity*, New Haven and London, 1980.

the infinite sequence, whose length, n, is greater than N, the probability that the relative frequency of heads for that segment is within d of 0.5 will be greater than $(1 - e)$. But this is perfectly compatible with its being the case that there is some positive number d such that there is, as a matter of fact, *no* initial segment, no matter how long, such that the relative frequency of heads is within d of 0.5. It is logically possible that the relative frequency of heads either fails to converge to any value at all, or converges to some value other than 0.5, even though the tossing situation is unbiased, and the likelihood of the coin's coming up heads is exactly 0.5.

The second approach that we need to consider appeals to the idea of propensities. The fundamental contention here concerns the logical form of statements expressing probabilistic laws. The suggestion is that the logically perspicuous way to express, for example, the claim that it is a law that the probability that something having property P has property Q is equal to k, is as the claim that it is nomologically true that everything with property P has propensity k for having property Q.[2]

That contention can, in turn, be combined with different views of what it is that makes a generalization nomologically true. If one accepted the simple regularity theory, its being a law that the probability that something with property P has property Q is equal to k would merely be a matter of its being true without exception that if something has property P, it also has propensity k for having property Q. If, instead, one accepted the view advanced above, it would be a matter of a certain contingent relation between property P, and the property of having propensity k of having property Q.

Since propensities are thought of as actual properties of actual objects, this account, in contrast to the limiting-frequency approach, does take seriously the problem of providing truth-makers for statements expressing statistical laws. It does, however, involve at least two serious and fundamental difficulties. The first concerns the question of what explanation can be offered of talk about propensities. That some explanation is needed is surely clear. In the first place, propensities are conceptually related both to corresponding occurrent properties and to corresponding

---

[2] Compare Railton, op. cit., p. 218.

dispositional properties. Thus, for example, the property of having a 0.5 propensity for disintegrating within 12 days is conceptually related to the occurrent property of being a case of disintegration. The property of having a 0.7 propensity for dissolving if placed in water is conceptually related to the dispositional property of being water-soluble. One needs an account which explains the nature of these relations.

Secondly, if talk about propensities were treated as primitive and unanalysable, one would be left with an infinite number of primitive predicates, since there is, for example, an unlimited number of values that can be substituted for n in the expression 'has a $0.n$ propensity for disintegrating'.

Thirdly, there are logical relations between different propensities. If an atom has a 0.5 propensity for disintegrating within the next 12 days, it cannot also have a 0.7 propensity for disintegrating within that time. If talk about propensities were taken as primitive, these relations would be completely mysterious. One needs an analysis of propensity-talk from which such interrelations will drop out as consequences.

The natural response to this request for an explanation of talk about propensities is that the concept of a propensity can be viewed as a generalization of the concept of a disposition, in which both deterministic and indeterministic dispositions are included.[3] This is, in itself, a plausible view to take; however, it cannot be combined with the view that it is facts about propensities that serve as the truth-makers for statements expressing probabilistic laws. For as was argued in section 2.2, one of the central objections to an account of laws that brings in reference to dispositions and powers is that expressions referring to dispositions and powers involve intensional contexts, and truth-makers must be described in purely extensional terms. If the concept of a propensity is viewed as a generalization of the concept of a disposition, any employment of propensities in giving truth conditions for probabilistic laws will be open to the same objection.

This leads immediately to a second serious objection to any approach which invokes propensities. For regardless of whether

---

[3] See Popper's discussion, 'Suppes's Criticism of the Propensity Interpretation of Probability and Quantum Mechanics', *The Philosophy of Karl Popper*, ed. Schilpp, La Salle, Ill., 1974, pp. 1125–39, esp. pp. 1129–30; and Mellor's discussion in *The Matter of Chance*, Cambridge, 1971, esp. pp. 66–70, and 169 ff.

one views them as generalized dispositions, the fact is that, unless one treats expressions describing propensities as primitive terms, devoid of structure, such expressions do involve intensional contexts. Consider, for example, the property of having a 0.8 propensity for dissolving within 5 seconds in water where less than 10 per cent of the molecules involve heavy hydrogen atoms. Now it is logically possible that all samples of water that exist at any time are such that they contain less than 10 per cent heavy water molecules. If propensity talk were extensional, it would then be true, in such a world, that something would have a 0.8 propensity for dissolving within 5 seconds in water where less than 10 per cent of the molecules involve heavy hydrogen if and only if it had also had a 0.8 propensity for dissolving in water within 5 seconds. Since it is clear that this need not be the case in a world of the sort described, it follows that propensity talk is not extensional. Therefore no adequate account of the truth-makers for statements expressing probabilistic laws can involve reference to states of affairs whose characterization involves propensity-concepts.

If these conclusions are correct, then neither relative frequencies nor propensities can be appealed to in order to provide truth-makers for statements expressing statistical laws. It therefore seems reasonable to conclude that no facts concerning *particulars* can serve as the truth-makers for such statements.[4]

The suggested conclusion, then, is the same as for statements expressing non-probabilistic laws: the truth-makers must consist of, or at least involve, irreducible relations among universals. In the case of non-statistical laws, the main support for this conclusion involved an appeal to the special case of laws which fail to have instances of all possible types. That consideration has been appealed to here, but I have also tried to show that there is a problem about the truth-maker for a statement expressing *any* statistical law, and one which cannot be solved in terms of facts about particulars.

---

[4] For a somewhat different line of argument that seems to point towards a very closely related conclusion, see Lewis's discussion of the question of whether chances can be supervenient upon particular facts, in 'A Subjectivist's Guide to Objective Chance', pp. 290–2.

4.2 PROBABILISTIC LAWS AND RELATIONS AMONG UNIVERSALS

If relations among universals are the truth-makers for statements expressing probabilistic laws, the problem becomes one of specifying the relations in question. Given that we have an account of the truth-makers for statements expressing non-probabilistic laws, the natural way of proceeding is to attempt to modify that account to get one that is applicable to statements expressing probabilistic laws.

In the case of non-statistical laws, the basic idea was that the relations in question could be characterized as those contingent, irreducible, theoretical relations among universals which logically necessitate certain corresponding, universally quantified statements about particulars. How might this idea be modified to handle the case of statistical laws? Perhaps the two most natural possibilities are these:

(1) The reference to universally quantified statements about particulars might be replaced by reference to statements of an appropriately different sort.

(2) The relation of logical necessitation might be replaced by some other, weaker relation.

In evaluating the relative merits of these two possibilities, it will be sufficient, I believe, to consider only one especially simple type of statistical law, namely, that expressed by statements of the following form:

> It is a law that the probability that something with the occurrent property P has the occurrent property Q is equal to k.

The corresponding relation among universals will be referred to as the relation of *probabilification to degree k*.

Let us now consider the first possibility. When the relation of nomic necessitation holds between properties P and Q, the statement that is logically necessitated is: For all x, if x has property P, then x has property Q. What statement about particulars should this be replaced by if one is to have an account of probabilification to degree k? Three possibilities immediately suggest themselves:

(*a*) The proportion of things with property *P* which have property *Q* is equal to *k*.
(*b*) In any infinite sequence of randomly selected things with property *P* the limit of the relative frequency with which those things possess property *Q* is equal to *k*.
(*c*) For all *x*, if *x* has property *P*, then *x* has a *k* propensity of having property *Q*.

In view of the discussion in the preceding section, however, it would seem that none of these possibilities can provide a satisfactory account of the truth-makers for statements expressing probabilistic laws. The first two are excluded because neither is entailed by the statement expressing the corresponding law. Its being a law that the probability that something with property *P* has property *Q* is equal to *k* does not entail that the proportion of things with property *P* which have property *Q* is equal to *k*. Nor, as we have also seen, does it entail that the limit of the appropriate relative frequency in an infinite sequence of randomly selected things with property *P* will be equal to *k*.

The third statement, in contrast, is entailed by the statistical law. But we have seen that the notion of a propensity, which it involves, is itself a non-extensional notion, and therefore cannot be employed in specifying the relations among universals which serve as the truth-makers for statements expressing statistical laws.

Let us consider, then, alternative (2), namely, replacing the relation of logical necessitation by some weaker relation. What might this relation be? The obvious choice is that of logical probability, or degree of confirmation. How will the analysis run if this relation is selected? A natural suggestion is that it will have the following general form:

Probabilification to degree *k* is that contingent, irreducible, theoretical relation between universals which is such that its holding between properties *P* and *Q* makes it probable to a degree *d* that *S* is the case.

Assume that the analysis should take this form. One then needs to specify what *S* and *d* are. Given that we are attempting to modify the account offered for non-statistical laws, *S* should be some statement about particulars, and one which involves concepts related to the universals *P* and *Q*. But precisely what statement?

And what is the value of d? In particular, is it equal to k or not? And if it is not, how does the number k enter into the analysis?

First, then, the problem of choosing an appropriate S. One possibility is to select as S the type of statement that enters into an account of laws of the nomic necessitation sort. However this does not seem very promising. For suppose that it is a law that the probability that something with property P has property Q is equal to k. How likely is it, given this, that everything with property P has property Q? The answer is not at all clear. If the world contains an infinite number of objects, the likelihood is infinitesimal. If it contains only a finite number of objects, the likelihood will depend upon just how many things there are.

The natural way of attempting to circumvent this problem is by instead selecting, as S, some probabilistic statement. As in the entailment analysis considered above, there are three main candidates:

> (a) For all x, if x has property P, then x has a k propensity of having property Q.
> (b) The proportion of things with property P which have property Q is equal to k.
> (c) In any infinite sequence of randomly selected things with property P the limit of the relative frequency with which those things possess property Q is equal to k.

The first alternative has to be rejected for reasons already mentioned. The shift from the relation of entailment to that of confirmation does not affect the objections raised.

If the second suggestion were accepted, one would have an account of the following form:

> Probabilification to degree k is that contingent, irreducible, theoretical relation between universals which is such that its holding between properties P and Q makes it probable to degree d that the proportion of things with property P which have property Q is equal to k.

The problem with this analysis is that it is not at all clear what the value of d should be. Suppose, for example, that k is some irrational number, such as $\pi/4$. If the world contains only a finite number of things, then the probability that a proportion k of them have property Q will be zero. So unless one is prepared to accept the not very plausible view that statistical laws require not merely instances, but an infinite number of them, the present approach will not be able to provide a satisfactory account of probabilistic laws in which k is irrational, since it will not be able to distinguish between possible laws that differ only with respect to the specific irrational number which is involved.

A way of avoiding this difficulty is by adopting the third suggestion, and thus offering the following account:

> Probabilification to degree k is that contingent, irreducible, theoretical relation between universals which is such that its holding between properties P and Q makes it probable to degree d that in any infinite sequence of randomly selected things with property P, the limit of the relative frequency with which those things possess property Q is equal to k.

By explicitly introducing reference to infinite sequences, this third approach avoids the problem just considered. At the same time, it gives rise to a different difficulty which is no less serious. For given that there may very well not be such infinite sequences of actually existent individuals of the relevant sorts, what account can one offer of the truth conditions of statements about the limits of relative frequencies? It would seem that one must appeal either to propensities, or to possible individuals, or to statistical laws. The first two appeals violate the requirements of extensionality; the third would render the analysis circular.

At this point the attempt to extend the analysis from the case of non-probabilistic laws to that of probabilistic ones by replacing the relation of entailment by that of logical probability, appears to be foundering. But one should recall a question which was raised when this general approach was suggested, namely, the question of whether d might not be equal to k. None of the possibilities explored to this point have had that feature. It might be helpful, then, to attempt to construct an account which does.

The idea, then, is to attempt to construct an account having the following logical form:

> Probabilification to degree k is that contingent, irreducible, theoretical relation between universals such that its holding between properties P and Q makes it probable, *to degree k*, that S is the case.

What might S be? The most plausible choice, I think, is not a sentence, but the following open formula: 'Any x that has property P also has property Q'. This gives one the following analysis:

> Probabilification to degree k is that contingent, irreducible, theoretical relation between universals such that its holding between properties P and Q makes it probable, to degree k, that any particular x that has property P also has property Q.

This account looks appealing. Upon reflection, however, it can be seen to be unsatisfactory. The reason emerges if one asks how the expression 'any particular x that has property P also has property Q' is to be interpreted. A natural answer is that it is to be taken as equivalent to the open material conditional, 'if x has property P, then x has property Q'. But while this choice is a natural one, the resulting account is unsatisfactory. One way of seeing this is as follows. Any adequate analysis must be such that its being the case that it is a law that the probability that something with property P has property Q is equal to k, together with its being the case that x has property P, makes it probable, to degree k, that x has property Q. But this does not follow from the proposed analysis. The reason is that since x can satisfy the open *material* conditional 'if x has property P, then x has property Q' if and only if it satisfies the logically equivalent open formula, 'either x lacks property P and x has property Q', it follows that a state of affairs which precluded anything's having property Q, but made it probable, to degree k, that any given thing would lack property P, would make it probable, to degree k, that any given x would satisfy the material conditional, 'if x has property P, then x has property Q'. Yet the fact that such a state of affairs obtained, together with the fact that x has property P, would not make it probable, to degree k, that x has property Q. The problem, in short, is that statistical laws of the sort we are considering license what might be called 'probabilistic inferences' of the following form:

> It is a law that the probability that something with property P has property Q is equal to k, and x has property P. Therefore is follows (with probability k) that x has property Q.

and such inferences would not be justified on the present account.

Another way of seeing that the analysis must be defective turns upon the point that contraposition does not hold for statistical laws. That is to say, its being a law that the probability that something with property P has property Q is equal to k does not entail that it is a law that the probability that something which lacks property Q also lacks property P is equal to k. Indeed, it does not even follow that there is any law dealing with the probability that something which lacks property Q will lack property P. If the above analysis were correct, however, contraposition would hold for statistical laws. For anything which makes it probable, to degree k, that any given x will satisfy the open formula 'if x has property P, then x has property Q', must also make it probable, and to the same degree k, that any given x will satisfy the open formula 'if x lacks property Q, then x lacks property P', since logical probabilities remain unchanged under the substitution of logically equivalent expressions.

One does not get a satisfactory account, then, if one construes the expression 'any particular x that has property P also has property Q' as equivalent to the material conditional 'if x has property P, then x also has property Q'. At this point it might be suggested that one needs to shift to some stronger type of conditional. It seems to me, however, that such a move is unpromising. For on the one hand, it would seem that nothing weaker than a subjunctive conditional will enable one to escape the difficulties. And on the other, if one does employ subjunctive conditionals, the account would seem to be circular given that reference to laws is apparently needed in providing truth-makers for subjunctive conditional statements.

It seems to me that one needs to modify one's conception of the logical form of the desired account. The two objections considered above point, however, in the direction of the required modification. The thrust of the first objection, for example, is that the definition, offered above, of the relation of probabilification to degree k, is such that its holding between two properties P and Q makes it likely, to degree k, that something either lacks property P or possesses property Q, whereas the relation one wants to capture is concerned rather with the likelihood that something will have property Q, *given that it has property P*. This suggests the following revision:

Probabilification to degree k is that contingent, irreducible, theoretical relation between universals such that its holding

> between properties P and Q, together with its being the case
> that x has property P, makes it probable, to degree k, that x
> has property Q.

This modification seems intuitively plausible, and it also handles the problem raised by the second objection. For given that it does not follow from the fact that the logical probability of s given r and t is equal to k that the logical probability of not-t given r and not-s is equal to k, neither will it follow from the fact that the relation of probabilification to degree k holds between properties P and Q that, if there were negative properties, not-Q and not-P would stand in the relation of probabilification to degree k.

So far so good. However, at least one problem remains. Suppose that it is a statistical law that the probability that a radon atom will undergo decay in a ten-day period is equal to 0.837. What is the probability that a radon atom, having a kinetic energy equal to the mean kinetic energy of radon atoms found in a gas at a temperature of 400 °C, will decay within a ten-day period? It would seem that this must also be equal to 0.837. For if, on the contrary, the probability of decay increased to, say, 0.894 when the temperature reached 400 °C, this would surely imply that it was not a statistical law that the probability of a radon atom's decaying within ten days was equal to 0.837, and that what one had, instead, were a number of statistical laws, corresponding to different temperatures. But how does this bear upon the above definition of probabilification to degree k? The answer is that the above definition does not have the desired consequence. For suppose that the relation of probabilification to degree 0.837 holds between the property of being a radon atom and the property of decaying within a ten-day period. Given that fact together with the fact that something is a radon atom, the probability that it will decay within ten days will be equal to 0.837. But the above definition of probabilification to degree k does not imply that given those facts, together with the additional fact that the radon atom in question has a certain kinetic energy, that the probability that it will decay within ten days is equal to 0.837.

In short, statistical laws have the following important feature: they give rise to probabilities that cannot, so to speak, be overridden before the fact. The probabilities can, of course, be overridden after the fact. If there is a single radon atom in a sealed

container, and reliable observers maintain, ten days later, that it is
still there, the probability that it decayed during the ten-day period
is very, very close to zero, rather than being equal to 0.837. But no
facts concerning earlier states of affairs can serve to make the
probability different from 0.837.

What one needs to capture then, is the idea of a state of affairs
such that its holding between two properties P and Q, together
with the fact that x has property P, in some sense 'determines', or
'fixes', the probability that x has property Q. This idea would
have been captured automatically if it had been possible to
characterize statistical nomological relations as ones whose holding
logically entails some corresponding probability statement. Unfortunately, we have seen that that approach is not viable. The
question, therefore, is how to capture this idea of statistical laws
'determining', or 'fixing' probabilities, when the relation of logical
entailment is replaced by that of logical probability.

Suppose that it is a statistical law, L, that the probability that x
has property Q at time t, given that x has property P at time t, is
equal to k. Let us say that a state of affairs, S, is admissible at time
t, relative to L, if and only if the probability that x has Q at time t,
given that x has P at time t, together with L and S obtaining, is
equal to k. The problem, then, is to find some general characterization of admissible states of affairs.

There are two intuitions concerning admissible states of affairs
which, initially, seem very plausible. The first is that the
probability that x has Q at time t, given that x has P at time t, and
that it is a law that L, should not be affected by what states of
affairs obtain at times prior to t. The other is that the probability
should not be affected by what other, logically compatible laws
there may happen to be. This suggests the following view:

> S is a type of state of affairs that is admissible at time t,
> relative to L, if and only if S's obtaining is logically
> compatible with L's obtaining, and either (1) S involves only
> states of affairs obtaining at times prior to t, or (2) S consists
> of universals standing in nomological relations, either prob-
> abilistic or non-probabilistic, or (3) S consists of some
> combination of states of affairs of the preceding two types.[5]

[5] Compare Lewis's discussion of the very closely related problem of specifying
what he refers to as 'admissible propositions', ibid., pp. 272–6.

There is, however, an important objection to such an account, as Lewis points out.[6] What if a reliable precognizer comes to believe that x will have Q at time t? Would not such a state of affairs be one that, though temporally prior, bears upon the relevant probability?

Many philosophers would challenge the underlying assumption that precognition is logically possible. The issue is a difficult one, however, and rather than pursue it here, it seems best to offer an account that is neutral on this issue.

If one adopts, at least for the moment, the supposition that precognition is logically possible, the intuition concerning the irrelevance of prior facts needs to be reformulated as follows. The reason that temporally prior states of affairs are generally admissible is not that they are temporally prior to the state whose likelihood is in question; it is rather that they are not causally posterior to that state. So one needs to replace reference to temporally prior states by reference to states that are not causally posterior. But this change must also be accompanied by another, since the requirement that admissible states be temporally prior ones also rules out states that either logically involve, or logically preclude, the state whose likelihood is in question. Another clause must therefore be added to exclude these possibilities. The account which results from these two changes is as follows:

> S is a type of state of affairs that is admissible relative to individual x, time t, and law L.

if and only if

> S's obtaining is logically compatible with L's obtaining, and also logically compatible both with x's having the relevant property at time t, and with x's not having the relevant property at time t, and either (1) S involves only states of affairs that are not causally dependent upon x's having the relevant property at time t, or upon x's not having that property at time t, or (2) S consists of universals standing in nomological relations, either probabilistic or non-probabilistic, or (3) S consists of some combination of states of affairs of the preceding two types.

Two objections to this revised analysis need to be considered.

---

[6] Ibid., p. 274.

The first is that this account of admissible states of affairs lands one in a conceptual circle. For it involves the notion of causal dependence, and that in turn presupposes the concept of laws of nature, including probabilistic ones.

This objection would be undercut if one could establish that causal relations do not presuppose underlying laws. In Chapter 6, I shall be considering how causal relations and causal laws are connected, and I shall be offering arguments against the view that causal relations are reducible to causal laws plus the non-causal properties of, and relations between, events. Initially, those arguments might seem to provide strong support for the singularist view of causation—according to which it is possible for there to be causally related events that do not fall under causal laws. However the two views just mentioned do not exhaust the range of alternatives. It turns out that there is a third possibility, according to which causal relations presuppose causal laws, without being logically supervenient upon such laws together with the non-causal properties of, and relations between, states of affairs. Moreover, in Chapter 8, I shall attempt to show that that third alternative is preferable to the singularist conception. Accordingly, I do not believe that the present difficulty is to be answered by embracing a singularist account.

Should the objection then be viewed as grounds for setting aside Lewis's appeal to the idea of a reliable precognizer, and returning to the analysis of the concept of admissible states of affairs which is in terms of the idea of temporal priority? The problem with this response is that it presupposes that the concept of temporal priority does not itself involve any causal notions. If, on the contrary, the direction of time, for example, can only be analysed in terms of the direction of causation, the same circularity will threaten the attempt to explain admissible states of affairs in terms of temporal priority. And as will become clear later, especially in sections 7.1 and 9.3, I believe that there are reasons for taking seriously a causal analysis of temporal concepts. As a consequence, it does not seem to me that returning to the first analysis of the concept of an admissible state of affairs provides a satisfactory way out.

How, then, is one to respond to the above objection? The correct reply emerges, I suggest, if one considers the case of scientific theories. For there it is often the case that a number of theoretical terms may be implicitly defined by their place within a single theory. The situation, therefore, may be precisely the same with respect to the notions of probabilistic laws and causation. So if the concept of a probabilistic law involves that of admissible states of affairs, and if the latter involves, directly or indirectly, the notion of causal dependence, and if, finally, causal relations presuppose underlying laws, the conclusion to be drawn is simply that rather than there being independent conceptual theories for the notion of probabilistic laws and for that of causation, there must be a single theory which incorporates both concepts. Such a theory, in turn, could easily be constructed by combining the theory of probabilistic laws set out in this chapter with the approach to causation to be developed in Chapters 8 and 9. A Ramsey/Lewis approach could then be applied to that more complex theory, thereby enabling one to define all of the relevant notions. The threatened circularity can be handled, therefore, once it is appreciated that a number of theoretical terms may be introduced by a single theory.

The second objection points to another possible circularity which may be generated if the concept of admissible states of affairs involves the notion of causal dependence, and if the latter presupposes that of causal laws. For if this is the case, then whether L_1 is a probabilistic law would seem to depend upon what other probabilistic laws there are, since the latter will make a difference with respect to what states of affairs are admissible, at a given time, relative to L_1. But, then, might there not be a world, and a sequence of possible probabilistic laws—$L_1, L_2, \ldots L_i, \ldots L_n$—where (1) for each L_i other than L_n, L_i is a law only if L_{i+1} is a law, and (2) L_n is identical with L_1? And if there were such a sequence of laws, looping back on itself, would not the resulting mutual dependence mean that it would be indeterminate whether any of them were laws in the world in question?

The answer to this second objection is as follows. If L_i's being a law is to be dependent upon whether L_{i+1} is a law, then L_{i+1} must be a causal law, since otherwise it would have no bearing upon what states of affairs were admissible relative to L_i. If there is to be a vicious circle, therefore, all the laws in the loop must be causal laws. It will be shown, however, in section 8.4, and in the Appendix, that causal loops are impossible. As a consequence, the suggested circularity cannot arise.

In any case, it does not appear to be crucial in the present context to resolve the question of precisely what account of admissibility is correct. For it would seem that whatever account of admissible states turns out to be best can be used to construct an account of the relation of probabilification to degree k. If, for example, the account suggested above is correct, one can offer the following analysis of the relation of probabilification to degree k:

R is the relation of probabilification to degree k

if and only if

> R is a contingent, irreducible, theoretical relation between universals such that its holding between properties P and Q, together with its being the case that x has property P at time t, together with the obtaining of any state of affairs, S, which is admissible relative to x, time t, and P's standing in relation R to Q, makes it probable to degree k that x has Q at time $t-$ where S is admissible relative to individual x, time t and P's standing in relation R to Q, if and only if S's obtaining is logically compatible with P's standing in relation R to Q, and also logically compatible both with x's having Q at time t and with x's not having Q at time t, and either (1) S involves only states of affairs that are causally dependent neither upon x's having Q at time t, nor upon x's not having Q at time t, or (2) S consists of universals standing in some nomological relation, or (3) S consists of some combination of states of affairs of the preceding two types.

Statistical laws may come, of course, in a variety of forms. Indeed, it is easy to see that no a priori limit can be placed upon the number of probabilistic nomological relations that may be needed to provide truth-makers for statements expressing statistical laws. The relevant argument is just a variation of that employed in the case of non-statistical laws. Thus, let P, Q, and R be any three properties. There could be a statistical law to the effect that the probability that something with property P has either property Q or property R is equal to k. Moreover, this could be a basic law. Then, if it is true that there are no disjunctive properties, it follows that the relevant truth-maker will have to consist of a triadic relation among universals. And in general, there could be properties P, Q_1, Q_2, ... Q_{n-1} such that it was a basic statistical law that the probability that something with property P has either property Q_1 or property Q_2 or . . . property Q_{n-1} is equal to k. The truth-maker for a statement expressing this law would involve an n-adic relation among universals. The number of possible probabilistic nomological relations is therefore in principle unlimited.

These additional probabilistic nomological relations do not appear, however, to generate any conceptual problems. It would seem to be a straightforward matter to develop a general account of probabilistic nomological relations that would cover not only the above relations, but those associated with statistical law-statements of every possible logical form. For the method of moving from the account of the dyadic relation of probabilification to degree k to a general account of probabilistic nomological relations will simply parallel the move, in the non-probabilistic case, from the relation of nomic necessitation to the general account of non-statistical nomological relations.

It appears, therefore, that the account offered, in the previous two chapters, of the truth-makers for statements expressing non-probabilistic laws can be modified, in a relatively straightforward and natural way, to provide one with an account of the truth-makers for statements expressing probabilistic laws. The main change required is to replace the relation of logical entailment by the relation of logical probability.

4.3 THE FORMAL PROPERTIES OF PROBABILISTIC LAWS

In section 3.4, in surveying some of the attractive features of the view that laws are relations among universals, I argued that one advantage of this view is that it makes it possible to show that non-probabilistic laws have the formal properties they are normally taken to have. In the present section I shall argue that the same is true in the case of probabilistic laws.

In order to do this, we need to have some system of axioms for the formal theory of statistical laws. I believe that a plausible way of generating such a system of axioms is by taking a set of axioms for probability theory, making whatever modifications are necessary, and then adding what may be referred to as the homogeneity postulate.

Consider, then, the following axioms for probability theory:

(A_1): Prob(B, A) is a single-valued function taking as its values nonstandard real numbers, and such that $0 \leq$ Prob(B, A) ≤ 1.

(A_2): If A is a subclass of B, Prob(B, A) $= 1$.

(A_3): If B and C are mutually exclusive,
Prob($(B \cup C)$, A) $=$ Prob(B, A) $+$ Prob(C, A).

(A_4): Prob($(B \cap C)$, A) $=$ Prob(B,A) $\times$ Prob(C,$(B \cap A)$).[7]

The next step is to modify these statements so that we have ones that are true of probabilistic laws. Consider A_1: it says that for any two classes, A and B, there is some unique number, r, between zero and one, which represents the likelihood that something belongs to B given that it belongs to A. This statement must be weakened substantially if the result is to be a statement that is true of statistical laws, since it is not true that for any two classes, A and B, there is a corresponding statistical law. What one will have, instead, is a statement that is conditional in form, and which rather than asserting that there are statistical laws corresponding to any two such classes, merely formulates a condition that any such law must satisfy:

(S_1): If Law-Stat(B, A, k), then $0 \leq k \leq 1$, and for all j such that $j \neq k$, it is not the case that Law-Stat(B, A, j)

where, for the moment, the expression 'Law-Stat(B, A, k)' is to be interpreted as equivalent to the statement that it is a statistical law that the probability that something is a B, given that it is an A, is equal to k.

There is one point that requires comment. Can it be a statistical law that the probability that an A is a B is equal to zero? Or equal to one? If probabilities are restricted to the standard reals, this is possible. But if infinitesimals are admitted, it is not. So it would seem that in (S_1), $0 \leq k \leq 1$ should be replaced by $0 < k < 1$. We

[7] These are essentially the axioms used by Salmon in *The Foundations of Scientific Inference*, Pittsburgh, 1966, p. 59, which are, in turn, based upon Reichenbach's discussion in *The Theory of Probability*, Berkeley, Calif., 1949, sects. 12–14. The notation used for conditional probabilities is, however, the more common one, rather than that preferred by Salmon and Reichenbach, and I have incorporated the idea that it is better if the probability function be allowed to take infinitesimals as values, rather than only standard real numbers.

shall see shortly, however, that it is convenient to reinterpret Law-Stat(B, A, k) slightly, and when this is done, (S_1) will be correct as it stands.

Next (A_2): even if infinitesimals were not being employed, it would certainly not be the case that whenever A is a subclass of B, it is a statistical law that the probability that an A is a B is equal to one, since it may be merely accidentally true that A is a subclass of B. Nor will it do to try to avoid this problem by strengthening the antecedent so that it asserts that it is nomologically necessary that A is a subclass of B, for then one will not have a probabilistic law.

The most natural response to this difficulty is simply to drop (A_2). Doing that, however, would necessitate complicating some of the other axioms. This can be avoided if one simply reinterprets Law-Stat(B, A, k) as: If k is not equal to one or to zero, then it is a statistical law that the probability than an A is a B is equal to k; if k is equal to one, then it is either logically necessary or nomologically necessary that anything which is an A is a B; if k is equal to zero, then it is either logically necessary or nomologically necessary that nothing is both an A and a B. This reinterpretation admittedly has an artificial air, but it does not affect the truth of the earlier axiom, S_1, and it does make the overall formal development smoother than it would otherwise be, for reasons that will become clear shortly.

Given this modification, one can now adopt the following postulate:

> (S_2): If it is logically or nomologically necessary that A is a subclass of B, then Law-Stat(B, A, 1).

In the case of A_3, it appears necessary both to nomologically strengthen the existing antecedent, and to add a clause postulating the existence of at least two of the relevant statistical laws. When these two changes are made, one has the following axiom:

> (S_3): If B and C are either logically or nomologically mutually exclusive, and if there exist j, k, and m such that at least two of the following are the case—Law-Stat(B, A, j), Law-Stat(C, A, k), and Law-Stat(($B \cup C$), A, m)—then there must also exist j, k, and m such that all three are the case, where $m = j + k$.

Finally, (A_4): all that is necessary here is that one add a clause postulating the existence of at least two of the relevant statistical laws:

(S_4): If there exist j, k, and m such that at least two of the following are the case—Law-Stat(B, A, j), Law-Stat (C, ($B \cap A$), k), and Law-Stat (($B \cap C$), A, m)—then there must also exist j, k, and m such that all three are the case, where $m = j \times k$.

This completes the modifications of (A_1) to (A_4).[8] The final step involves adding some sort of *homogeneity postulate* asserting that if it is a statistical law that the probability that something is a B, given that it is an A, is equal to k, then it is also a statistical law that the probability that something is a B, given that it belongs to a relevantly admissible subclass of A, is also equal to k. The considerations that need to be taken into account in defining the notion of relevantly admissible subclasses are just those considered above in connection with the specification of admissible non-nomological states in the definition of the relation of probabilification to degree k. The following would therefore seem to be a reasonable formulation of the desired postulate:

(S_5): If it is the case that Law-Stat(B, A, k), and if (1) C is such that the fact that something is both an A and a C does not entail either that it is a B or that it is not a B, and (2) whether an A is a C does not causally depend either upon its being a B or upon its not being a B, then it must also be the case that Law-Stat(B, ($A \cap C$), k).

Given the view that it is certain relations among universals that are the truth-makers for statements expressing statistical laws, can it be shown that the above axioms for statistical laws turn out to be true? To do this, one would need a much more detailed account than I have offered, since the discussion here has been restricted to the special case of statistical laws that can be expressed by statements of the form 'It is a law that the probability that an A is a B is equal to k' and where there are unique properties by virtue of which something is an A or a B. Nevertheless, I think that this partial account will at least enable one to sketch a line of argument which will make plausible the general claim.

---

[8] I am indebted to John Collins for pointing out a flaw in my earlier formulations of (S_3) and (S_4).

First, postulate (S_1): the central question that needs to be considered here is whether it is possible to have distinct k and j, neither of which is equal to zero or to one, such that one has both Law-Stat(B, A, k) and Law-Stat(B, A, j), where something is an A by virtue of possessing property P, and a B by virtue of possessing property Q. Let R_k be the relation of probabilification to degree k, and R_j the relation of probabilification to degree j. Assume that it is possible to have both Law-Stat(B, A, k) and Law-Stat(B, A, j). Then one must have P and Q standing in relation R_k and also in relation R_j. But the definition offered earlier of the relation of probabilification to degree k asserts that the probability that a has Q, given that a has P, that P and Q stand in relation R_k, and that any admissible state of affairs S obtains, must be equal to k. Then, since any nomological relation among universals that is logically compatible with P and Q's standing in relation R_k is admissible, P and Q's standing in relation R_j is admissible, from which it follows that the probability that a has Q, given that a has P, that P and Q stand in relation R_k, and that P and Q stand in relation R_j, must be equal to k. A parallel argument, starting from the definition of probabilification to degree j, will also show that this probability must be equal to j. So k and j cannot be distinct.

The satisfaction of (S_2) is a trivial matter, since it in no way depends upon the account of the nature of statistical laws. It follows immediately from the reinterpretation assigned to Law-Stat(B, A, k) in the case where k is equal to one.

Next, (S_3): this axiom introduces a complication, since it involves statistical law-statements with different logical forms. This means that one should really introduce nomological relations other than those of probabilification to degree k, as defined above. The resulting gain, however, would be marginal, while the discussion would be considerably more complex. Consequently, it seems best to adopt the following, simplified approach. First, rather than introducing additional nomological relations, let us simply use the expression 'probabilification to degree k' in an extended sense, in which a statistical nomological relation will be said to be one of probabilification to degree k if it is the appropriate truth-making relation among universals for any statistical law-statement of the general form, 'It is a law that the probability that an A is a B is equal to k', regardless of the logical complexity of the terms A and B. Secondly, however complex the terms A, B, and C may be, let us speak about corresponding properties, P, Q, and T.

If we adopt these somewhat loose modes of expression, the argument can be put as follows. Suppose that P and Q stand in relation R_j, and P and T in relation R_k. In view of the definitions of these relations, the probability that something is a B, given that it is an A and the above relations obtain, must be equal to j, while the probability that something is a C, given that it is an A, and that the above relations obtain, must be equal to k. If it is logically impossible for something to be both a B and a C, it follows by virtue of the rules governing relations of logical probability that the probability that something is either a B or a C, given that it is an A, and that the above relations obtain, must be equal to $(j + k)$. Moreover, it seems clear that there could not be any admissible state of affairs, S, that would override this probability, since in order to do so it would have to override the probabilities generated either by the statistical law that the probability that an A is a B is equal to j, or by the statistical law that an A is a C is equal to k. Therefore it must be a statistical law that the probability that something is either a B or a C, given that it is an A, is equal to $(j + k)$.

If it is not logically, but only nomologically impossible for something to be both a B and a C, the argument needs to be reformulated slightly. In addition to the relation R_j between P and Q, and the relation R_k between P and T, one needs to refer to the relation of nomic exclusion between Q and T, and to consider probabilities relative to this more complex nomological state of affairs. But given that change, the argument goes through as before. In either case, then, it must be a statistical law that the probability that something is either a B or a C, given that it is an A, is equal to $(j + k)$.

Axiom (S_3) also asserts that if it is logically or nomologically impossible for something to be both a B and a C, and if it is a statistical law that the probability that an A is a B is equal to j, and a statistical law that the probability that something is either a B or a C, given that it is an A, is equal to m, then it must also be a statistical law that the probability that an A is a C is equal to k, where k is $(m - j)$. The proof of this would be very similar to the argument just set out.

Axiom (S_4) can be established in a very similar way. Suppose that P and Q stand in relation R_j, and that the conjunctive property, P & Q, stands in relation R_k to T. In view of the definitions of these relations, the probability that something is a B, given that it is an A, and that P & Q stand in relation R_j, and that the conjunctive property, P & Q, stands in relation R_k to T, must be equal to j, while the probability that something is a C, given that it is both an A and a B, and that the same nomological states obtain, must be equal to k. And from these two facts it follows, by virtue of confirmation theory, that the probability that something is both a B and a C, given that it is an A, and that the same nomological states obtain, must be equal to $(j \times k)$. Moreover, it seems clear that there could not be any admissible state of affairs, S, that would override this probability, since in order to do so it would have to override the probabilities generated either by the statistical law that the probability that an A is a B is equal to j, or by the statistical law that the probability that something that is both an A and a B is a C is equal to k. Therefore it must be a statistical law that the probability that something is both a B and a C, given that it is an A, is equal to $(j \times k)$.

This shows that the account offered does have the consequence that if it is a statistical law that the probability that an A is a B is equal to j, and a law that the probability that something is a C, given that it is both an A and a B, is equal to k, then it must also be a law that the probability that something is both a B and a C, given that it is an A, is equal to m, where m is $(j \times k)$. But postulate (S_4) involves two other assertions. First, that if it is a law that the probability that an A is a B is equal to j, and a law that the probability that an A is both a B and a C is equal to m, then it must also be a law that the probability that something that is both an A and a B is also a C is equal to k, where k is m/j. Second, that if it is a law that the probability that something that is both an A and a B is a C is equal to k, and a law that the probability that an A is both a B and a C is equal to m, then it must also be a law that the probability that an A is a B is equal to j, where j is m/k. However, these other two consequences of (S_4) could be established in similar fashion.

Finally, the homogeneity postulate, (S_5): this states that if it is a law that the probability that an A is a B is equal to k, then it will also be a law that the probability that something that is both an A and a C is a B is equal to k, provided that C satisfies certain conditions. In particular, it must not be the case that something's being both an *A* and a *C* entails that it is a *B*, nor precludes its being a *B*, and it must not be the case that something's being both an *A* and a *C* is causally dependent either upon its being a *B*, or upon its not being a *B*. These conditions ensure that (S_5) must be true given the account advanced above of the nature of statistical laws. For we saw that relations such as that of probabilification to degree *k* must be defined in such a way that the probabilities that they generate cannot be overridden by the addition of admissible states of affairs, and the account of admissibility was such that when *C* satisfies the above conditions, the fact that something is both an *A* and a *C* will be admissible relative to the statistical law in question. Therefore, the probability that something is a *B*, given that it is both an *A* and a *C*, must also be equal to *k*. To show that one has a statistical law one would also have to demonstrate that this probability cannot be overriden by the addition of admissible states of affairs. But that would seem to follow in a straightforward fashion from the fact that the probabilities to which the initial statistical law gives rise cannot be overridden by states that are admissible relative to it.

The view that statistical laws are to be identified with certain relations among universals appears to entail, then, postulates (S_1) to (S_5). It is possible, of course, that (S_1) to (S_5) do not provide a complete axiomatic basis for the derivation of all the formal properties of statistical laws. However, they surely constitute a central core, and the fact that the account of the truth-makers for statements expressing probabilistic laws that has been advanced here does entail the truth of (S_1) to (S_5) provides, I suggest, important support for the present approach.

4.4 SOME ADVANTAGES OF THE PRESENT APPROACH

I should like to conclude the discussion of probabilistic laws by mentioning some of the advantages of the present account. Most of these can be dealt with very briefly, since they involve points that have already been discussed in connection with non-probabilistic laws.

In the first place, then, this approach provides an account of the truth conditions of statements expressing statistical laws that does not involve, either implicitly or explicitly, any nomological concepts. No appeal has been made, for example, to subjunctive conditionals, or to implicitly nomological notions such as propensities.

Second, the account is free of any reference to possible worlds, or to hypothetical objects or events—in contrast to what appears to be required, for example, in the case of approaches that refer to the limits of relative frequencies in infinite sequences. The truth-makers for statements expressing statistical laws are, on the present account, facts about the actual world—although once again, as with non-probabilistic laws, they are facts about universals, rather than facts about particulars, and they are theoretical facts, rather than observable ones.

Third, the account provides a clear answer to the question of the difference between statements expressing statistical laws and statements of accidentally true statistical generalizations. The truth-makers for the former are facts about universals, and for the latter, facts about particulars.

Fourth, and in view of this difference between the truth-makers for the two sorts of statements, it becomes clear why statistical laws support subjunctive conditions in a way that mere generalizations do not.

Fifth, this account allows for the possibility of basic statistical laws which lack instances, provided that uninstantiated universals are possible.

Sixth, this approach to statistical laws provides a simple explanation of the apparent failure of extensionality within ordinary statements of statistical laws. The reason that substitution of co-extensive predicates within ordinary formulations of statistical laws may fail to preserve truth is that when those laws are reformulated in a philosophically adequate way, predicates having as their extensions collections of particulars are replaced by terms that refer instead to the universals associated with the previous predicates.

Seventh, this view of the nature of probabilistic laws provides a satisfactory account of the reason why one is justified in accepting certain hypotheses concerning such laws. Consider, for example, a hypothesis to the effect that there is some number k, between h and j, such that it is a law that the probability that something is a B, given that it is an A, is equal to k. Given this hypothesis, probabilities can be assigned to different possible relative frequencies—including not only the frequency with which something is a *B*, given that it is an *A*, but the frequency with which something is a *B*, given that it belongs to some admissible subclass of *A*. If there is good agreement between the observed relative frequencies and those that are likely upon the hypothesis in question, and if no alternative hypothesis generates such a close correspondence, then it is natural to suppose that this provides some justification for accepting the hypothesis. But what rationale is to be offered of this? The most natural view would seem to be that it is a matter of an inference to the explanation. But to adopt this view, one must advance an account of the nature of statistical laws which does not identify them with the facts that determine actual relative frequencies. The present approach satisfies this requirement, since laws are identified with relations among universals, whereas it is facts about particulars which determine actual relative frequencies. There is therefore no obstacle to explaining the latter in terms of the former.

The final attractive feature of the present account was discussed in the previous section: it appears to make possible a derivation of all the purely formal properties of probabilistic laws.

In conclusion, it appears that the view that the truth-makers for nomological statements are contingent, irreducible, theoretical relations among universals can be extended to the case of probabilistic laws. The account is more complicated than in the case of non-probabilistic laws. But on the other hand, the extension is a natural one, which does not appear to involve any insurmountable technical problems, and the resulting account of the truth-makers for statements expressing probabilistic laws has a number of very attractive features. In short, there would seem to be excellent grounds for accepting it.

PART III:

Causation

— 5 —

The Basic Issues

Given the account of laws set out above, we are now in a position to turn to the question of the nature of causation. This chapter contains a brief survey of some central issues that we shall need to consider.

Since a satisfactory explanation of the nature of causation must provide accounts both of causal laws and of causal relations between particular states of affairs, one of the crucial issues concerns the relation between, and the relative priority of, causal laws and causal relations.

As regards the latter issue, there would appear to be three views that might be adopted. The first is that causal laws are primary, and causal relations secondary. This basic thesis can be expressed in slightly different ways. One way of formulating it is in terms of the second thesis of Humean supervenience, set out in Chapter 1:

> The truth values of all singular causal statements are logically determined by the truth values of statements of causal laws, together with the truth values of non-causal statements about particulars.

A more common formulation, however is in terms of a thesis concerning the possibility of *analysing* statements about causal relations between particular states of affairs in terms of statements of causal laws, together with statements concerning the non-causal properties of, and relations between, states of affairs. This thesis can be expressed in various ways, but a typical formulation might run as follows:

> State of affairs *a* caused state of affairs *b*

can be analysed as

There are non-causal properties P and Q, and a non-causal relation R, such that a has property P, b is the only state of affairs with property Q standing in relation R to a, and it is a causal law that states of affairs with property P are always accompanied by states of affairs with property Q which stand in relation R to them.

A second and radically different position is what is known as the singularist view—which claims that it is causal relations between states of affairs that are primary, and causal laws that are secondary. This view is forcefully expressed by C. J. Ducasse in the following passage:

The supposition of recurrence is thus wholly irrelevant to the meaning of cause; that supposition is relevant only to the meaning of law. And recurrence becomes related at all to causation only when a law is considered which happens to be a generalization of facts themselves individually causal to begin with. A general proposition concerning such facts is, indeed, a causal law, but it is not causal because it is general. It is general, i.e., a law, only because it is about a class of resembling facts; and it is causal only because each of them already happens to be a causal fact individually and in its own right (instead of, as Hume would have it, by right of its co-membership with others in a class of pairs of successive events). The causal relation is essentially a relation between concrete individual events . . .[1]

Ducasse's exposition of this second view presupposes a Humean account of laws, according to which they are to be equated with certain sorts of regularities; however, the position can be combined with radically different accounts of the nature of laws. One could, for example, adopt the view advanced above, and maintain that the truth-makers for statements expressing causal laws are certain relationships among the relevant universals— including whatever universals are involved when states of affairs stand in causal relations.

A second point that deserves to be noted is that the singularist position is also compatible with very different views concerning whether causal relations can be analysed in terms of non-causal properties and relations. Ducasse maintained that such an analysis

[1] Ducasse, 'On the Nature and the Observability of the Causal Relation', in *Journal of Philosophy*, 23, 1926, pp. 57–67. Reprinted in *Causation and Conditionals*, ed. Sosa, Oxford, 1975, pp. 114–25. See p. 118.

is possible.[2] Other advocates of this second position have held that causal concepts must be taken as primitive.

If causal relations are essentially relations between particular states of affairs, then it must be logically possible to have causal relations for which there are no covering laws. Or even for there to be a world that is full of causally related events, but which is completely anomic. But are these things really possible? Most present-day philosophers, I believe, think not, and as a result, hold that causal relations must be supervenient upon causal laws together with non-casual facts.

There is, however, a third alternative to the supervenience view on the one hand, and the singularist view on the other. For one might hold, first, that causal relations *presuppose* underlying laws—that is, that it is impossible for two states of affairs to be causally related unless they fall under some causal law—but second, that whether two states of affairs are causally related need not always be logically determined by their non-causal properties and relations, together with what causal laws there are.

This third view has not found many advocates. In the next chapter, however, I shall attempt to show that the supervenience view of causal relations is exposed to serious objections. If I am right in thinking that those objections make it very difficult to accept the supervenience position, the third view might turn out to be very appealing, since it would then be the only viable alternative to a singularist view.

5.2 THE SUPERVENIENCE OF CAUSAL FACTS UPON NON-CAUSAL FACTS

A second important issue concerns the relation between causal facts and non-causal facts. Is it possible for causal facts to vary, so to speak, while all non-causal facts remain fixed? Or does the totality of non-causal facts logically determine what causal facts obtain?

If one accepts the supervenience view of causal relations, the question of whether causal facts are supervenient upon non-causal facts comes down to the question of the nature of causal laws. One

---

[2] Ibid., pp. 116, 120 ff.

could hold that causal law-statements are true by virtue of irreducible causal facts, and ones that do not involve, as constituents, irreducible causal relations holding between particular states of affairs. However, this position has not, I believe, been put forward by anyone. Moreover, at least given the account of laws advanced above, it is not a plausible view. For if laws are relations among universals, there can be special causal facts corresponding to causal law-statements only if some of the universals that are constituents of the law are universals of an irreducibly causal sort. The universals that enter into any first-order law must, however, be properties of, or relations among, particulars. Accordingly, there would have to be special causal facts involving particulars. A supervenience view of causal relations would, therefore, be precluded.

The nature of the non-causal facts upon which causal facts are supervenient will depend, if the supervenience view of causal relations is sound, upon the correct account of the nature of laws. If laws are simply cosmic regularities, it will follow that causal facts are supervenient upon non-causal facts about particulars. On the other hand, if the view of laws advanced here is correct, causal facts will be supervenient upon non-causal facts, some of which involve relations among universals.

The situation will be quite different if either the singularist account, or the third view, is correct. For then it would seem to be a completely open question whether causal facts are logically supervenient upon non-causal facts. Some proponents of a singularist view, for example—such as Ducasse—offer an account of causal relations and causal laws according to which they are supervenient upon non-causal facts. Other proponents of a singularist view—such as Elizabeth Anscombe—maintain that causal relations are not supervenient upon non-causal facts.[3]

5.3 THE ANALYSABILITY OF CAUSAL CONCEPTS

A third important issue is whether causal concepts are analysable in non-causal terms, or whether they must be treated as primitive. If causal facts are logically supervenient upon non-causal facts,

---

[3] Anscombe, 'Causality and Determination', ibid., pp. 63–81.

then it would seem that it must be in principle possible to analyse causal concepts in non-causal terms. But what if causal facts are not supervenient upon non-causal ones? I suspect that many philosophers would hold that it then follows that causal concepts cannot be analysed in non-causal terms. If this were right, I think that it would be a rather strong argument for the view that causal relations must be supervenient upon non-causal facts—since, as was argued in section 1.3, there seem to be good reasons for insisting that causal and nomological concepts must be analysable in terms of concepts that are neither causal nor nomological.

It seems clear, however, that the inference in question is unsound. For consider the corresponding argument in the case of theoretical terms. To be a realist with regard to theoretical entities is to hold that facts concerning the relevant unobservable entities (or events, etc.) are not logically supervenient upon observable facts. But, as we saw in section 1.2, this does not imply that it is impossible to offer an analysis of theoretical terms.

The situation regarding theoretical concepts is, moreover, directly pertinent to the case of causal concepts, if the account of causality to be set out below is correct. For one of the central conclusions for which I shall be arguing is that causal relations are *theoretical* relations, and that they are theoretical not merely with respect to what is observable, either in some broad or narrow sense, but with respect to *all* non-causal facts, both observable and unobservable.

This way of looking at causation has not, I believe, been seriously investigated. The assumption that causal relations are supervenient upon non-causal facts has, since the time of Hume, been a central feature in almost all accounts of the nature of causation. In Hume's own case, it is easy to understand the appeal of this assumption. Given his insistence that causal concepts stand in need of analysis, coupled with his theory of meaning—according to which all ideas must be capable of being traced back to impressions—the supervenience thesis is unavoidable. What is less clear, however, is why this assumption has continued to be so widely accepted, long after it has been generally recognized that Hume's version of an empiricist theory of meaning must be abandoned—given, for example, that it cannot provide a satisfactory account of the meaning of theoretical terms, realistically construed.

The feeling—expressed by Tom Beauchamp and Alexander Rosenberg, among others—appears to be that although Hume appeals, in his discussion of causation, to a certain theory of meaning, such an appeal is not essential. Thus, in their discussion of this issue, Beauchamp and Rosenberg refer to Anscombe's criticisms of Hume, and they suggest that her 'strategy is to attack Hume's epistemology, hoping that its inadequacies will cast a pall over his theory of causation'.[4] They contend that this strategy is misguided, on the grounds that Hume's conclusions do not depend in any crucial way upon his semantic and/or epistemological assumptions. I believe that they are mistaken in this matter, and that, in particular, once Hume's theory of meaning is jettisoned, there is no satisfactory argument for the claim that causal facts are supervenient upon non-causal facts. This, in turn, opens the door for the type of theory that I shall be defending—according to which, first, causal relations are theoretical relations, realistically conceived; second, they are theoretical relative to all non-causal facts, but; third, causal concepts are nevertheless analysable in non-causal terms, by means of whatever techniques are required for the analysis of theoretical concepts in general.

5.4 CAUSAL ORDER, PRIORITY, AND EFFICACY

One of the most striking differences between fundamental causal relations and nomological ones, is that the former are necessarily asymmetric, while the latter are not. Suppose that it is a law that anything with property P also has property Q. This would seem to be perfectly compatible with its also being a law that anything with property Q has property P. For if this were not possible, it would be impossible for one sort of condition to be both causally necessary and causally sufficient for another. If having property P is causally sufficient for having property Q, it must be a law that whatever has property P has property Q. Similarly, if having P is causally necessary for having Q, it must be a law that whatever has Q also has P. So if having property P is both causally necessary and causally sufficient for having Q, it must be a law that something has property P if and only if it has property Q. The

---

[4] Beauchamp and Rosenberg, *Hume and the Problem of Causation*, New York and Oxford, 1981, p. 82.

relation of nomic necessitation cannot, therefore, be necessarily asymmetric. By contrast, if having property P is a causally sufficient condition for having property Q, then having property Q cannot be a causally sufficient condition for having property P. Causal necessitation is necessarily asymmetric.

The same is true if one considers causal relations between particular states of affairs. If state of affairs a causes state of affairs b, it cannot be the case that b causes a. The relation of causation is also necessarily asymmetric.

Fundamental causal relations differ from nomological ones, then, in that the former are necessarily asymmetric, while the latter are not. Another, closely related point is this: if R is some fundamental causal relation, and S is the corresponding ancestral relation,[5] then S is also necessarily asymmetric. One of the fundamental issues, therefore, is what account is to be given of the necessary asymmetry of both causal relations and their ancestrals.

Another central aspect of causal relations is that they define an ordering of *causal priority*. This is not necessarily captured by an account of causal asymmetry, for the following reason. Given that the ancestral of any relation is necessarily transitive, and since any relation that is both asymmetric and transitive defines a strict (partial) ordering of the relevant entities, any account of the asymmetry of causal relations will show why causal relations, through their ancestrals, give rise to a strict partial ordering of states of affairs. But if R is any asymmetric and transitive relation, then the inverse relation, R^*, is also asymmetric and transitive, and it generates an ordering that is indistinguishable from that generated by R, except that it is opposite in direction. This means that any satisfactory account of causal priority, in addition to explaining the asymmetry of causal relations, must also supply some account of why it is the *direction* defined by those relations, rather than that defined by the inverse relations, that is *the* direction of causal processes.

Closely related to the idea that causes are prior to their effects is

[5] Intuitively, the ancestral of a relation R is a relation S that stands in the same relation to R as the relation of being an ancestor of stands to the relation of being a parent of. More precisely, relation S is the ancestral of relation R if and only if, for any x and y, x and y stand in relation S if and only if x belongs to the set $M(y)$, where $M(y)$ is the smallest set that satisfies the following two conditions: (1) $M(y)$ contains y; (2) for any z, if $M(y)$ contains z, then it also contains every w such that w stands in relation R to z.

the idea that they are efficacious, that they in some sense 'produce', or 'bring about' their effects. Thus, one question that one can ask about any account of causation, and especially of the rationale that it offers for selecting one direction as the direction of causal priority, is whether the account enables one to make sense of the difficult notion of *causal efficacy*. I believe that one of the virtues of the account to be offered here is that it does enable one to do this.

5.5 CAUSATION AND TIME

Another crucial issue concerns the relation between causal concepts and temporal concepts. Hume, in his famous account of causation, took the view that temporal priority is one of the notions that enters into the analysis of the relation of cause and effect: '. . . we may define a cause to be *an object followed by another, and where all the objects, similar to the first, are followed by objects similar to the second*'.[6]

A number of philosophers have followed Hume on this matter. Patrick Suppes, for example, in his formal presentation of a probabilistic theory of causation, explicitly takes the direction of time as given, and defines the direction of causal processes in terms of it.[7]

Analyses of causation that employ temporal concepts typically have to face two familiar objections. The first arises out of constraints that such approaches impose upon causal relations. Usually they imply that it is logically impossible, first, for cause and effect to be simultaneous, and second, for a cause to occur after its effect. Both of these theses, and especially the first, have been seriously questioned. Any account of causation that involves temporal notions must therefore show either that it does not have such implications, or that these consequences are in fact acceptable.

The second objection concerns whether one is running the analysis in the proper direction if one employs temporal concepts in the analysis of causal ones. For many philosophers have held

[6] Hume, *An Inquiry Concerning Human Understanding*, pt. II, sect. VII. Compare Hume's *A Treatise of Human Nature*, Bk. I, pt. III, sect. II.
[7] Patrick Suppes, *A Probabilistic Theory of Causality*, Amsterdam, 1970, p. 12.

that causal concepts are more basic than temporal ones, and have attempted to analyse the latter in terms of the former.

The question whether a satisfactory account of causal priority can be given in terms of temporal priority will be taken up in the first two sections of Chapter 7. Then, in Chapters 8 and 9, I shall offer an account of causation that does not involve temporal concepts. Finally, in section 9.3, I shall briefly discuss the prospects for a causal theory of time.

5.6 CAUSATION AND SPATIOTEMPORAL CONTINUITY

Another important issue is whether genuine action at a distance—either spatial, or temporal, or both—is logically possible. Can there be gappy causal processes, or must all causal processes exhibit spatiotemporal continuity? Several philosophers seem to hold that it is logically necessary that causal processes be spatiotemporally continuous. This view was advanced by Hume, though he entertained some doubt on the matter,[8] and it plays a central role in the analysis of causation proposed by Ducasse.[9] More recently, Wesley Salmon, in his discussion of causal explanation, seems to maintain that causal processes must exhibit spatiotemporal continuity.[10] This issue will be taken up in Chapter 7, in connection with the discussion of Salmon's account of causation.

5.7 CAUSAL RELATIONS AND PROBABILISTIC LAWS

A final important issue concerns the possibility of causal relations falling under laws that are merely probabilistic. Some philosophers have expressed serious doubts concerning this possibility. Hugh Mellor, for example, says that he 'can make no sense of explicating the cause–effect relation in terms of less than

[8] Hume, *A Treatise of Human Nature*, Bk. I, pt. III, sect. II.
[9] Ducasse, op. cit., p. 116.
[10] Salmon, 'Theoretical Explanation', *Explanation*, ed. Korner, Oxford, 1975, pp. 118–43. See pp. 127–8. 'Why Ask "Why?"?', *Proceedings and Addresses of the American Philosophical Association*, 51/6, 1978, pp. 683–705. See pp. 689 ff.

100 per cent correlation'.[11] On the other hand, a number of philosophers—such as Hans Reichenbach, Wesley Salmon, and Patrick Suppes—have argued very strongly in support of the desirability of an approach to causation that explicitly allows for the possibility of such probabilistic causal relations.

There are, I think, two issues that need to be disentangled. The first is whether it makes sense to speak of one state of affairs as causing another when the underlying law connecting the relevant properties is a probabilistic law. Just how difficult this question is depends upon the correct view of the relation between causal laws and causal relations. If the singularist account is correct—so that causal relations are primary, and causal laws secondary—the issue is unproblematic. Given that there can, on that view, be causal relations without any covering laws, there will surely be no problem about causal relations when the relevant laws are merely probabilistic.

Suppose, on the other hand, that the singularist view is mistaken, and that causal laws, not causal relations, are primary. The issue will then be much less clear-cut. The correct attack upon the problem in that case will presumably be to examine the concept of a non-probabilistic causal law, to see what account can be offered, and then to determine whether the analysis can be modified, so as to generate a satisfactory account of probabilistic causal laws.

The second issue is whether, if it does make sense to speak of causal relations where the underlying law is merely probabilistic, the relation between the states of affairs is different than in cases falling under non-probabilistic laws.

In the case of this second issue, too, the answer is less clear-cut if the singularist view is mistaken. If the supervenience view is correct, and if, in addition, causal laws may be either probabilistic or non-probabilistic, then I think that it is plausible to hold that what one has is not exactly a single causal relation, but a family of similar relations, one associated with non-probabilistic causal laws, and the others corresponding to each of the different probabilities that can enter into probabilistic causal laws. If, on the other hand, causal relations presuppose underlying laws, but supervenience does not obtain, then it may well turn out that causation is a single relation.

[11] Mellor, 'Comment', in Korner, op. cit., pp. 146–52. See p. 147.

— 6 —

Laws and Causal Relations

How are causal relations between particular states of affairs related to causal laws? In the preceding chapter, I suggested that there are three main views that can be taken on this matter. The choice among those alternatives would seem, moreover, to be crucial for any account of causation. In spite of that fact, the question of which view is correct has been all but totally neglected in present-day discussion of causation. Indeed, since the time of Hume one answer has more or less completely dominated philosophical thinking about causation. In this chapter I shall attempt to show that the view in question is in fact profoundly problematic, and probably has to be abandoned.

6.1 THE THREE ALTERNATIVES

6.1.1 *The Supervenience View*

According to the first view, which is currently accepted by the vast majority of philosophers, causal laws are primary, and causal relations secondary. There are different ways of attempting to make this claim more precise. One way of doing so is in terms of the second thesis of Humean supervenience, as set out in Chapter 1:

> *The Thesis of the Humean Supervenience of Causal Relations*
>
> The truth values of all singular causal statements are logically determined by the truth values of statements of causal laws, together with the truth values of non-causal statements about particulars.

This formulation seems perfectly satisfactory, and I believe that it has the merit of being the most modest statement of the basic claim. It says, in effect, only that it is in principle possible to analyse singular causal statements in terms of non-causal statements together with statements of causal laws. It does not indicate, even in outline, the form that such an analysis would take.

But on the other hand, if this claim is to be rendered plausible, it seems likely that one needs at least a sketch of how such an analysis might run. As a consequence, a more common approach is to put forward the general supervenience claim by advancing some specific account of how statements concerning causal relations between particular states of affairs can be analysed in terms of statements of causal laws together with non-causal statements. A typical suggestion is this:

State of affairs a caused state of affairs b

means the same as

There are non-causal properties P and Q, and a non-causal relation R, such that a has property P, b has property Q, b is the only state of affairs with property Q standing in relation R to a, and it is a causal law that any state of affairs with property P is always accompanied by a state of affairs with property Q which stands in relation R to it.

This analysis has the defect, however, of implying that the underlying laws must be non-probabilistic. It might seem that this defect is easily corrected: simply replace the clause 'it is a causal law that any state of affairs with property P is always accompanied by a state of affairs with property Q which stands in relation R to it' by something like 'there is some number p such that it is a causal law that for any state of affairs with property P, the probability that there is some state of affairs with property Q which stands in relation R to it is equal to p'. The resulting analysis only works, however, in the case of *immediate* causation. For suppose that the reason that it is a causal law that any state of affairs with property P is accompanied, with probability p, by a state of affairs with property Q which stands in relation R to it, is that there are the following two causal laws:

A state of affairs with property P is accompanied with probability p_1, by a state of affairs with property S which stands in relation R_1 to it;

A state of affairs with property S is accompanied with

probability p_2, by a state of affairs with property Q which stands in relation R_2 to it;

where p is equal to $(p_1 \times p_2)$, where R is the logical product of R_1 and R_2, and where there are no other relevant laws. If this were so, one could have a case where there is a state of affairs with property P that does not give rise to one with property S, and so does not give rise to one with property Q, but where something else causes there to be a state of affairs with property Q which just happens to stand in relation R to the state of affairs with property P. The above analysis would then imply the false claim that the state of affairs with property P caused the state of affairs with property Q that stands in relation R to it.

It might seem, however, that this defect also could easily be corrected. One could view the above analysis as giving an account of direct or immediate causation, and then define indirect causation as the ancestral of that relation. To say that state of affairs a causes state of affairs b would thus be to say that either a is the immediate cause of b, or that there is some chain of events, $c_1, c_2, \ldots c_i, \ldots c_n$ such that a is the immediate cause of c_1, c_n is the immediate cause of b, and each c_i (other than c_n) is the immediate cause of c_{i+1}, the next element in the causal chain.

But this approach involves an assumption that can be seen to be unacceptable. For it is being assumed that if a is the cause of b by means of some intervening causal process, then there must be some state of affairs, c_1, that is causally intermediate between a and b, and that is such that there is *no* state of affairs that is causally intermediate between a and c_1. But why should this be the case? Why might it not be the case that all causal processes are infinitely divisible, so that for any two causally related states of affairs, there are causal intermediaries? As a consequence, it would seem that the relation of causation cannot be analysed in terms of that of immediate causation.

It is not, therefore, a trivial matter to set out an adequate, and completely general analysis of singular causal statements, even given the notion of a causal law. Fortunately, the difficulties involved are not germane to the present issue. Accordingly, it will be simplest, and sufficient, to work with a formulation of the supervenience thesis that, rather than offering an analysis of singular causal statements, is along the more modest lines of the first formulation set out above.

It will be useful, however, to offer a slightly more explicit version of that formulation, and also to distinguish between the cases of immediate causation and mediate causation.

The Second Thesis of Humean Supervenience: Immediate Causation

There are meaning postulates for causal expressions such that a statement of the form, 'State of affairs a is the *immediate* cause of state of affairs b', is true if and only if that statement is entailed by the set of statements that consists of:

(1) the relevant meaning postulates;
(2) all true statements of causal laws;
(3) all true statements concerning the non-causal, non-relational properties of states of affairs a and b;
(4) all true statements concerning non-causal relations between states of affairs a and b.

The Second Thesis of Humean Supervenience: Mediate Causation

There are meaning postulates for causal expressions such that a statement of the form, 'State of affairs a is the *mediate* cause of state of affairs b', is true if and only if there is some non-empty, and possibly uncountably infinite set, S, of states of affairs, other than a and b, such that the above statement is entailed by the set of statements which consists of:

(1) the relevant meaning postulates;
(2) all true statements of causal laws;
(3) all true statements concerning the non-causal, non-relational properties of states of affairs a, b, and those belonging to set S;
(4) all true statements concerning the non-causal relations among states of affairs a, b, and those belonging to set S.

The introduction of the reference to what may be an uncountably infinite set, S, enables one to deal with the case where a is the cause of b, but not the immediate cause, and where, moreover, there is no chain of intervening states of affairs that are tied together by the relation of immediate causation.

Many, and perhaps most, philosophers who accept the above thesis of the Humean supervenience of causal relations would also accept the following closely related claim concerning causal laws:

The Thesis of the Humean Supervenience of Causal Laws

The truth values of all statements of causal laws are logically determined by the truth values of all non-causal, non-nomological statements about particulars.

Moreover, when the latter thesis is accepted, and combined with the former, one is led to the following, stronger claim regarding the supervenience of causal relations:

The Strong Thesis of the Humean Supervenience of Causal Relations

The truth values of singular causal statements are logically determined by the truth values of non-causal, non-nomological statements about particulars.

We shall not, however, be concerned with either of these theses in the present chapter. The discussion of laws in Part II has already provided strong grounds for rejecting the first of these claims, and this in turn eliminates one rather plausible line of argument in favour of the second claim. None the less, it may still turn out to be the case that causal facts concerning particulars are logically supervenient upon non-causal facts. That issue cannot be decided until we have resolved the question to be discussed in the present chapter, namely, that of the relation between causal laws and causal relations between particulars.

6.1.2 *The Singularist View*

An alternative view is that causal relations between states of affairs are primary, and causal laws secondary. As was noted in the previous chapter, such is the view of C. J. Ducasse: 'The causal relation is essentially a relation between concrete individual events . . .'[1] It also appears to be the position of Elizabeth Anscombe, though her essay 'Causality and Determination' is admittedly not entirely explicit on this matter. Consider, for example, the following passage from very near the end of her essay:

---

[1] Ducasse, 'On the Nature and the Observability of the Causal Relation', p. 118.

188 *Causation*

Meanwhile in non-experimental philosophy it is clear enough what are the dogmatic slumbers of the day. It is over and over again assumed that any singular causal proposition implies a universal statement running 'Always when this, then that'; often assumed that true singular causal statements are derived from such 'inductively believed' universalities. Examples indeed are recalcitrant, but that does not seem to disturb. Even a philosopher acute enough to be conscious of this, such as Davidson, will say, without offering any reason at all for saying it, that a singular causal statement implies *that there is* such a true universal proposition—though perhaps we can never have knowledge of it. Such a thesis needs some reason for believing it![2]

In this passage Anscombe is explicitly calling into question only the proposition that causal relations presuppose underlying *non-probabilistic* laws. Moreover, the discussion throughout her essay is focused upon the claims, first, that causal relations must be instances of exceptionless generalizations, and second, that causes must necessitate their effects, and someone who holds that there can be probabilistic causal laws could agree that those claims should be rejected, without thereby being forced to conclude that causal relations do not presuppose *any* laws—either probabilistic or non-probabilistic.

None the less, I think it is reasonably clear that Anscombe does wish to accept the singularist view. For consider one of her central lines of argument. It rests upon the following intuition concerning the nature of causality:

. . . causality consists in the derivativeness of an effect from its causes. This is the core, the common feature, of causality in its various kinds. Effects derive from, arise out of, come of, their causes.[3]

Given this intuition, she goes on to say:

Now analysis in terms of necessity or universality does not tell us of this derivatedness of the effect; rather it forgets about that. For the necessity will be that of laws of nature; through it *we* shall be able to derive knowledge of the effect from knowledge of the cause, or vice versa, but that does not show us the cause as source of the effect. Causation, then, is not to be identified with necessitation.

If *A* comes from *B*, this does not imply that every *A*-like thing comes from some *B*-like thing or set-up or that every *B*-like thing or set-up has

___
[2] Anscombe, 'Causality and Determination', *Causation and Conditionals*, Sosa (ed.), Oxford, 1975, pp. 63–81. See p. 81.
[3] Ibid., p. 67.

an *A*-like thing coming from it; or that given *B*, *A* had to come from it, or that given *A*, there had to be *B* for it to come from. Any of these may be true, but if any is, that will be an additional fact, not comprised in *A*'s coming from *B*.[4]

As an argument, this does not seem to have much force, because everything turns upon the notion of an effect's deriving from, arising out of, its cause, and no analysis of this crucial notion is offered. My point here, however, is simply that the above line of thought dues strongly suggest a singularist view of causation. For why does Anscombe maintain that *A*'s coming from *B* does not imply either that every *A*-like thing comes from some *B*-like thing or set-up or that every *B*-like thing or set-up has an *A*-like thing coming from it? The answer, I think, is that Anscombe believes that causation is just a relation between concrete individuals, such as *A* and *B*, rather than, as Hume thought, something involving at least an implicit reference to corresponding *types* of individuals. But if this view is right, then *A*'s coming from *B* equally fails to imply either that there is some probability *p* that any *A*-like thing comes from some *B*-like thing, or that there is some probability *p* that a *B*-like thing will give rise to an *A*-like thing. Singular causal statements will not imply the existence of any laws, either probabilistic or non-probabilistic.

6.1.3 *A Third Alternative*

At first glance, the supervenience view and the singularist view might seem to exhaust the alternatives. But as we saw earlier, this is not the case. An advocate of the supervenience view affirms what a proponent of the singularist view denies: that all causal connections between states of affairs presuppose underlying causal laws. However, the supervenience view involves a further claim, to the effect that what causal relations there are is logically determined by the causal laws, together with the totality of non-causal facts. This means that it is possible to reject both the singularist view and the supervenience view. Perhaps singular causal facts are not logically determined by causal laws together with non-causal facts but, none the less, it is impossible for there to

---

[4] Ibid., pp. 67–8.

be singular causal facts for which there are no corresponding causal laws.

I am not aware of any philosopher who has advanced this third view. This may suggest that the third alternative is, perhaps, prima facie implausible. It does seem true that the nature of the relation between causal laws and causal relations is, at least initially, somewhat more puzzling on this third view than on either the supervenience view or the singularist view. For if one finds plausible the idea that causal relations presuppose causal laws, it is natural to be drawn to the clear account of that connection that is provided by the supervenience view. But on the other hand, if it turns out, as I shall attempt to show, that the supervenience view is exposed to very serious objections, this third possibility may then be an important alternative to the singularist position.

6.2 TWO RELATED ARGUMENTS AGAINST THE SUPERVENIENCE VIEW

In this section, and the next two, I shall set out four arguments against the view that causal relations between states of affairs are logically supervenient upon causal laws plus non-causal facts. Two of the arguments are very closely related, having essentially the same logical structure. They will be dealt with in the present section.

6.2.1 *The Causal Theory of Time*

The first argument concerns the possibility of a causal theory of time. The basic thrust of the argument is that it is logically possible that the universe might be in precisely the same state at different times, and that this possibility, when conjoined with the thesis of the Humean supervenience of causal relations, leads to the conclusion that a causal analysis of temporal concepts is logically impossible.

The argument can be stated as follows. Let P_1, M_1, and M_2 be complete temporal slices of the universe—either instantaneous or extended—where P_1 occurs before M_1, but not before M_2, and where M_1 and M_2 are qualitatively indistinguishable. (It might be a case of a deterministic universe where conditions are such that there is an endless repetition of every type of state that exists at any time. Alternatively, it might be an indeterministic universe in which, by accident, the state of the universe at one time happens to be qualitatively indistinguishable from its state at some other time.) The question now is this: can one give an account of temporal concepts, in causal terms, that allows it to be the case that P_1 is before M_1, but not before M_2, if the second thesis of Humean supervenience is true?

The argument is simplest in the case where P_1 is the *immediate* cause of M_1, so let us begin by considering that case. Given the thesis of Humean supervenience, it can be the case that P_1 immediately causes M_1 only if that statement is entailed by the set of statements—call it T—that contains:

(1) the relevant meaning postulates;
(2) all true statements of causal laws;
(3) all true statements concerning the non-causal, non-relational properties of P_1 and M_1;
(4) all true statements concerning the non-causal relations between P_1 and M_1.

Next, what must be the case if P_1 is not to be the immediate cause of M_2 as well? Given the thesis of Humean supervenience, the answer is that the statement that P_1 immediately causes M_2 must not be entailed by the set of statements—call it U—that contains:

(1) the relevant meaning postulates;
(2) all true statements of causal laws;
(3) all true statements concerning the non-causal, non-relational properties of P_1 and M_2;
(4) all true statements concerning the non-causal relations between P_1 and M_2.

How can it be the case that T entails the statement that P_1 immediately causes M_1, whereas U does not entail the statement that P_1 causes M_2? For this to be the case, T must contain some statement about M_1, such that U does not contain the corresponding statement about M_2. This statement cannot be one of the meaning postulates, nor one of the causal laws, since neither sort of statement, we can assume, will contain any reference to particular temporal slices. Nor can the difference be a matter of statements attributing non-causal, non-relational properties to M_1 and M_2 respectively, since they are, by hypothesis, qualitatively indistinguishable. The only possibility, therefore, is that there is some non-causal relation that holds between P_1 and M_1, but not between P_1 and M_2.

What might this relation be? Relations may be classified, following a traditional distinction, into internal and external ones—an internal relation being one that holds by virtue of the nature, or non-relational properties, of the relata.[5] Since M_1 and M_2 do not differ with respect to any of their non-causal, non-relational properties, there cannot be any internal, non-causal, relation which holds between P_1 and M_1, but not between P_1 and M_2. The only possibility, therefore, is an external relation.

What external relations can obtain between different, complete temporal slices of the world? There would seem to be at most three sorts: (1) temporal relations; (2) causal relations; (3) nomological relations. Then, since the second possibility is excluded, it follows that there must be either some temporal relation or some nomological relation which holds between P_1 and M_1, but not between P_1 and M_2.

It would however, seem implausible to accept the thesis of the Humean supervenience of causal relations without also accepting the parallel thesis with respect to nomological relations:

> *The Thesis of the Humean Supervenience of Nomological Relations*
>
> The truth values of statements concerning instances of laws of nature are logically determined by the truth values of nomological statements, together with those of non-nomological statements about particulars.

But given this thesis, one can draw the corresponding conclusion that P_1 can stand in some nomological relation to M_1 in which it does not stand to M_2 only if there is either some temporal relation, or some causal relation, which holds between P_1 and M_1, but not between P_1 and M_2.

From this it follows that P_1 can be the immediate cause of M_1, but not of M_2, only if there is some temporal relation that holds between P_1 and M_1, but not between P_1 and M_2.

---

[5] Armstrong, *Universals and Scientific Realism*, ii, Cambridge, 1978, p. 172.

But now suppose that the thesis of the Humean supervenience of causal relations is true. If M_1 is causally dependent upon P_1, while M_2 is not, there must be some non-causal fact about M_1, or its relation to P_1, such that there is no corresponding fact about M_2, or its relation to P_1. It may seem that such a fact is readily at hand. For is it not true that the end of the one temporal part P_1 has the same location as the beginning of M_1, and a different location from the beginning of M_2? A problem arises, however, when one asks what meaning is to be assigned to statements concerning the locations of objects at different times. What does it mean to say that the end of P_1 has the same location as the beginning of M_1?

Suppose that one adopts a relational view of space. Then possible coordinate systems at a given time will have to be defined in terms of objects existing at that time. But in order to make sense of comparisons of the locations of objects at different times one has to be able to explain what it is for coordinate systems existing at different times to be identical. Now if all the coordinate systems existing at a given time were qualitatively distinguishable by virtue of their relations to different sorts of objects, one might try to solve the problem by appealing to the idea that a requirement of continuity over time can serve to determine when coordinate systems existing at different times are temporal parts of a single enduring coordinate system. But even if this approach will work in some cases, it is not available in the case of the simple universe we are considering here, since, for every coordinate system that exists at a given time, there is another one, with axes that are differently oriented, that is qualitatively indistinguishable from it. As a consequence, the only way of sorting out which of a pair of qualitatively indistinguishable coordinate systems at a given time is the same coordinate system as some coordinate system existing at an earlier time is by being able to identify the individuals existing at the different times. This, however, lands one in a vicious circle. Identity of elementary entities over time rests upon causal relations, and these in turn must be grounded upon non-causal facts, if the thesis of the Humean supervenience of causal relations is true. Moreover, we have seen that the non-causal facts in question must include ones concerning the locations of different temporal parts of the relevant elementary entities. But if a relational view of space is correct, then, in the universe containing just the two neutrons, comparisons of locations at different times presuppose the identification of objects over time.

The upshot is that the possibility of rotationally symmetrical universes shows that one cannot accept both the Humean supervenience of causal relations, and a relational view of space. Suppose, then, that one adopts an absolute view. Is the problem resolved? It seems clear that it is not. To say that the end of P_1 has the same location as the beginning of M_1 will be to say that the end of P_1 is located in the same region as the beginning of M_1. But one now needs some account of what makes it the case that two regions existing at different times are identical. It does not seem acceptable to treat this relation as primitive and analysable.

Yet if it is to be analysed, what alternative is there to appealing to causal relations? But then, if the identification of spatial regions over time rests upon causal relations, one can consider a universe in which absolute space exhibits rotational symmetry, and then simply repeat the above argument—applying it to regions of absolute space, rather than to physical objects. Thus, let Q_1 be the region of space containing the end of P_1, and let N_1 and N_2 be the regions of space containing, respectively, the beginning of M_1 and the beginning of M_2. If Q_1 is to be the same region of absolute space as N_1, but a different region from N_2, then it must be the case that there is some causal relation that obtains between Q_1 and N_1, but not between Q_1 and N_2. If the thesis of the Humean supervenience of causal relations is true, this causal difference must be grounded in some non-causal difference between N_1 and N_2. If one were dealing with an absolute space that was not rotationally symmetrical about the relevant axis, there might be no problem about the existence of such a non-causal difference. But if one assumes that the universe is one that possesses the relevant rotational symmetries, in what could such a non-causal difference between N_1 and N_2 consist? There does not appear to be any satisfactory answer to this question.

If this is correct, we have arrived at the following conclusion. Regardless of whether a relational view of space or an absolute view is correct, the thesis that causal relations are supervenient implies that either the existence of rotationally symmetrical universes is impossible, or the relation of identity over time cannot be analysed in causal terms.

To sum up, I have argued in this section that the view that causal relations between states of affairs are logically supervenient upon causal laws together with non-causal facts has two consequences that seem unwelcome, and that therefore constitute at least prima facie objections to that view. The first is that it is impossible to give a causal analysis of temporal concepts. The second is that either the existence of rotationally symmetrical universes is logically impossible, or else it is impossible to give an account, in causal terms, of the identification over time of either physical objects or regions of space.

6.3 A THIRD ARGUMENT: THE CASE OF INVERTED UNIVERSES

The first argument discussed above was directed to showing that the supervenience view precludes offering an analysis of temporal concepts in causal terms. In this section I shall set out an argument for a closely related conclusion—to the effect that if the supervenience view is correct, it is impossible to explain the direction of time in terms of the direction of causal processes.

The argument may be put as follows: imagine that our world was actually created by a cosmic Laplace in the year two billion BC, and that its laws are both completely deterministic and symmetrical with respect to time. Before creating our universe, however, the Laplacean-style deity calculated that, given the intended laws and initial distribution of matter and energy, the universe would collapse completely in four billion years. Then, having noted the position and velocity that every particle would have just before that final collapse, at the very instant that he created this universe he created another universe a great distance away, with the same laws, but otherwise inverted with respect to this one. That is to say, the initial relative positions of particles in the other universe correspond exactly to the relative positions of particles in this universe just before the final collapse, but all the velocities are reversed.

Since the two universes have the same laws, and those laws are deterministic, and symmetrical with respect to time, the result will be a universe just like ours but running, so to speak, in reverse. Thus, for example, a temporal cross-section of our own universe at some point in AD 2000 will be just like a cross-section of the other universe at a corresponding point in 2000 BC, except that all the velocities will be reversed.

Let A_1 and A_2 be complete temporal slices of our universe in AD 2000 and AD 2001 respectively, and let B_1 and B_2 be the corresponding slices of the inverted universe, from 2000 BC and 2001 BC respectively. A_1 causally gives rise to A_2, but B_1 does not give rise to B_2, The question now is whether it is possible for this to be the case if the supervenience view is true.

If causal relations are supervenient, the above causal difference between the two universes can exist only if there is also a difference either with respect to causal laws, or with respect to non-causal facts concerning the relevant states of affairs. By hypothesis, the laws are the same in the two universes. So the difference must be with respect to non-causal facts.

What differences are there, aside from explicitly causal ones? The only other ones are with respect to the direction of time, and with respect to properties and relations that depend upon time. Thus, while A_1 is temporally prior to A_2, B_1 occurs after B_2, rather than before. And as a consequence, A_1 is not qualitatively indistinguishable from B_1, nor A_2 from B_2. For while the relative positions of the particles in A_1 agree exactly with the relative positions of those in B_1, and do so regardless of whether the slices are instantaneous, or have some temporal thickness, all the velocities are reversed.

But are such differences really non-causal? The answer will depend upon what account is given of the direction of time. If the direction of time is to be defined in terms of the direction of causal processes, then the above differences will not be non-causal. This in turn will imply that if Humean supervenience does hold for causal relations, then the world described above cannot really be logically possible, since it could not be the case that A_1 gives rise to A_2, while B_1 does not give rise to B_2. Yet surely the world described above is logically possible. Hence, if the direction of time is to be defined in terms of the direction of causation, the supervenience view of causal relations must be rejected.

The difference between this argument and the first argument, set out above, is perhaps worth underlining, and it may be put as follows. Some philosophers hold that while a complete analysis of all temporal concepts in causal terms is impossible, causal concepts do play a role in the analysis of at least some temporal notions. In particular, one might hold, first, that there is a concept of temporal betweenness, which does not have any direction associated with it, and which cannot be analysed in causal terms, but second, that this concept must then be combined with that of the direction of causation, to give analyses of those temporal concepts that involve a reference to the direction of time. For a philosopher who adopted this view, the third argument would have force while the first would not.

6.4 A FOURTH ARGUMENT: INDETERMINISTIC LAWS

In this section I want to offer a final, somewhat more direct argument against the supervenience view. The argument turns upon certain possibilities of causal relations in worlds that are partly indeterministic. One way of stating the argument involves probabilistic causal laws, and may be put as follows: suppose that it is a causal law that whenever there is a state of affairs, x, with property P, this always gives rise to a state of affairs, y, with property Q, where y is located in a small region of space that stands either in relation to R to x, or in relation S to x, the probability of each outcome being 50 per cent. Now given certain choices of property P, and relations R and S, it will be possible for there to be two states of affairs with property P, say x_1 and x_2, such that a region stands in relation R to x_1 if and only if it also stands in relation R to x_2, and similarly for relation S. If, for example, P were the property of having a certain triangular shape, and relations R and S were defined in terms of distances along the extensions of specified sides, Figure 1 would provide an example of the sort of situation that I have in mind. In the situation depicted, there is a y_1 that has property Q and that stands in relation R both to x_1 and to x_2, and similarly there is a y_2 that has property Q and that stands in relation S both to x_1 and to x_2. How should such a situation be described in causal terms, given the probabilistic law stated earlier? One possibility is this:

x_1's having property P caused y_1 to have property Q, and
x_2's having property P caused y_2 to have property Q.

But the correct description might equally well be:

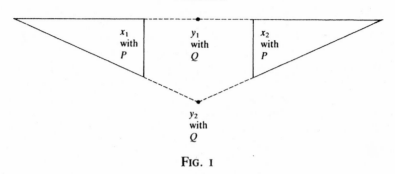

FIG. I

x_1's having property P caused y_2 to have property Q, and
x_2's having property P caused y_1 to have property Q.

So the problem is this: what would make one of these things the case, rather than the other?

One answer is that there might be an intervening causal process linking x_1's having P to y_1's having Q, and similarly, an intervening causal process linking x_2's having P to y_2's having Q, but none linking x_1's having P to y_2's having Q, or x_2's having P to y_1's having Q. If this were so, the ontological situation would be sorted out. But why need this be the case? Might not the probabilistic law in question be one describing *direct* causal processes? In the case just set out, that would involve causal connections over gaps. Some philosophers believe that the idea of a gappy causal process is ultimately incoherent. I shall argue, in the next chapter, that that view is mistaken. But in any case, the example can easily be reformulated to avoid that problem, as David Armstrong has pointed out. For example, one can imagine that one is dealing instead with causal laws that say:

> For any x, x's having property P will give rise either to x's having property Q or to x's having property R;
> For any x, x's having property S will give rise either to x's having property Q or to x's having property R.

Imagine, then, that a has both property P and property S, and also both property Q and property R. Does a's having P cause it to have property Q? Or is this caused by a's having property S? Here there is no possibility of appealing to an intervening causal process to sort out the ontological situation.

The upshot is that if the thesis of the Humean supervenience of causal relations is true, it follows that in this situation is cannot be the case either that *a*'s having property *P*, or that *a*'s having property *S*, caused it to have property *Q*. One must rest content with its being the case *only* that *a*'s having property *P* caused it to be the case that *a* had either property *Q* or property *R*, and similarly for *a*'s having property *S*. So if one asks 'What *would* have been the case if *a had* had property *P*, but not property *S*?' the answer is that there is no truth of the matter. For there is nothing that could possibly serve as the truth-maker for a counterfactual such as 'If *a* had had property *P*, but not property *S*, then *a* would have had property *Q*, but not property *R*.'

How might a proponent of the supervenience view respond to this argument? There seem to be three possibilities. First, one might argue that the notion of indeterministic laws—upon which the argument rests—is in the final analysis incoherent. Second, one might attempt to show that even if there can be indeterministic laws, there cannot be causal ones—perhaps on the ground that the notion of cause is only intelligible if it can be unpacked in terms of the notion of a condition that is causally sufficient to ensure that a certain type of effect will occur. Third, one might contend that there is nothing unacceptable about the 'ontological indeterminacy' that, if causal relations are supervenient, is involved in the above situation.

I have already tried to rule out the first response by offering, in Chapter 4, an account of probabilistic laws. But even if that account turned out to be unsatisfactory, it would be very surprising, in view of the nature of quantum physics, if there were no satisfactory account of probabilistic laws. Secondly, it is important to notice that the argument does not need the notion of probabilistic laws. As the second formulation illustrates, all that is required is the notion of indeterministic laws. Scepticism regarding probability or probabilistic laws is therefore not sufficient. One must also offer reasons for rejecting non-probabilistic, indeterministic laws of the sort mentioned in Chapter 2.

The second response is perhaps slightly more promising, since as was noted in Chapter 5, some philosophers have argued that there are serious difficulties attached to the notion of probabilistic causal relations, and I believe that they would hold that the same is true with regard to any indeterministic causal relations. But here, too, quantum physics appears to constitute a serious objection, since it certainly seems to contain laws that are both causal and probabilistic. In any case, I shall attempt to show, in Chapter 9, that what might be referred to as probabilistic causal relations are in fact perfectly intelligible.

What about the third response? It is not clear, perhaps, that ontological indeterminacy of the sort in question is unacceptable. But it does seem clear that, other things being equal, one would prefer an account that did not give rise to such ontological indeterminacy. The crucial question, therefore, is whether one can set out an intelligible account of causation that will enable one to avoid the indeterminacy. If one can offer such an account, and it is not otherwise problematic, then one has a good reason for rejecting the supervenience view. In Chapter 8, I shall attempt to show that such an account is available.

6.5 THE THIRD ALTERNATIVE

I have outlined four arguments against the view that causal relations are supervenient upon causal laws together with non-causal facts. I believe that those arguments constitute a very strong case against the supervenience view. If that is so, which of the remaining views should be adopted? The singularist view? Or the view that while causal relations presuppose underlying causal laws, what causal relations obtain is not necessarily determined by the causal laws together with the relevant non-causal facts?

The latter alternative should now have, I believe, a definite appeal, since it allows one to admit the force of the above arguments against the supervenience view, while enabling one to avoid some of the serious difficulties to which a singularist view is exposed. For, first of all, on a singularist view one is committed to the idea, which may seem rather unpalatable, that there could be a world full of causally related events yet utterly devoid of any laws. Secondly, and yet more serious, there is the very difficult problem of attempting to explain how causation, if it is nothing more than a relation between two concrete events, differs from the non-causal relation that obtains between two events when one merely happens to accompany the other. The third alternative, with its insistence that causal relations presuppose underlying laws, avoids both of these problems.

But the third alternative is not without its own problems. On the one hand, it is claimed that underlying laws are necessary if there are to be causal relations. On the other, that laws together with relevant non-causal facts are not always sufficient to determine what causal relations obtain. The combination of these two claims may appear to generated a puzzle. Given the second claim, it would seem that a proponent of this third view must hold that there is some special relation, *C*, that obtains between any two causally related states of affairs. The question, then, is what account is to be given of the first claim. Why are underlying laws necessary if there are to be causal relations? The answer will depend upon whether it is logically possible for there to be states of affairs that stand in relation *C*, but that do not fall under any law. If this is possible, the reason that underlying laws are necessary if there are to be causal relations might seem to be a very superficial one: our use of causal terms happens to be such that we do not say that two states of affairs are causally related unless it is the case both that they stand in the special relation *C* and that they fall under some law of the appropriate form. The difference between the third position and the singularist position might thus seem to be merely verbal: a proponent of the singularist approach says that two states of affairs are causally related if they stand in the special relation *C*; an advocate of the third view says that they are causally related only if they also fall under some law.

The alternative is to maintain that it is logically impossible for two states of affairs to stand in relation *C* unless they fall under some law. This gives one a philosophically distinctive position. But now one is confronted with the problem of explaining how a relation between first-order states of affairs can possibly entail the existence of some law under which those states of affairs fall. This is puzzling on any view of the nature of laws; however it is perhaps especially so if, as I have argued, laws are relations among universals.

6.6 SUMMING UP

Where does this leave us? What does seem clear is that the supervenience view, which has dominated philosophical thinking about causation since the time of Hume, must be abandoned. But which of the two remaining views should be adopted? Neither view is without its difficulties. If one opts for the singularist view, one is confronted with the possibility, which many would judge counter-intuitive, of worlds full of causally related events, but devoid of laws, and with the yet more serious problem of explaining how causation, so conceived, differs from the non-causal relation that obtains between two events when one merely happens to accompany the other. On the other hand, if one opts for the third view, one is confronted with the problem of explaining the relation between the standing of states of affairs in the special relation, C, and the existence of an appropriate underlying law.

The choice between these two accounts will be considered in Chapter 8. There I shall attempt to show that there is an extremely plausible general approach to the analysis of causal concepts that can be used to formulate both a supervenience view, and an account of the third, intermediate sort, but which cannot be used to formulate a singularist account. As a result, I believe that there is good reason for thinking that the singularist view of causation *is* unintelligible.

I shall also argue that a clear and acceptable account can be given of the relation that obtains, on the third view, between causal laws, and causal relations between states of affairs. The conclusion, therefore, is that while the problem of providing an analysis of causal concepts does eliminate the singularist view, one is still left with an intelligible alternative to the supervenience account. The arguments just advanced show that it is that alternative which should be accepted.

The Direction and Asymmetry
of Causation

A central task confronting any attempt to give an account of the nature of causation concerns what may be referred to as the problem of causal priority. As was noted in section 5.4, this problem involves two main aspects. First, both fundamental causal relations and their ancestrals are asymmetric. What account is to be given of the source of this asymmetry? Second, given this asymmetry, both fundamental causal relations and their ancestrals serve to partially order states of affairs. There are, however, two different directions associated with any ordering. One needs, therefore, to specify which direction is the direction of causation, and to explain why that direction has the significance it does.

This chapter contains a survey of the solutions that have been proposed. I shall attempt to show that none of the proposals is ultimately satisfactory. In doing so, however, I shall also attempt to show that an awareness of the difficulties that must be overcome points to certain conditions that must be met by any adequate account.

7.1 TEMPORAL PRIORITY

One very natural approach to the problem of causal priority involves the following two claims. First, that one state of affairs can be causally prior to another only if it is also temporally prior. Second, that it is precisely this fact that accounts for the asymmetry of fundamental causal relations and their ancestrals, and for the direction of causation. The relation of cause and effect is asymmetric because the relation of being earlier than is asymmetric, and the direction of causal processes is nothing more than the direction of time.

This appears to have been Hume's view, and it has also been

accepted by a number of present-day philosophers, such as Pat Suppes.[1] It is, however, exposed to at least three objections. The thrust of the first is that this approach leaves one without any satisfactory account of temporal priority. For if the direction of causal processes is to be analysed in terms of the direction of time, and the asymmetry of causation explained in terms of the asymmetry of temporal priority, then a causal theory of the direction of time is, of course, precluded, on pain of circularity. So an advocate of the present approach must hold either that the concept of temporal priority is analytically basic, or that it can be analysed in non-causal terms. However, it can be argued that neither of these is ultimately plausible.

As those who analyse the direction of causation in terms of the direction of time rarely go on to discuss the concept of temporal priority, it seems likely that they generally believe that the latter concept does not stand in need of analysis, that it can appropriately be taken as analytically basic. But I think that there is a strong argument against this view. It turns upon the fact that the relation of temporal priority has certain formal properties: it is necessarily irreflexive, necessarily transitive, and necessarily asymmetric. For given a proposition that seems to express a necessary truth—such as that no event is earlier than itself—some explanation of the necessity is surely desirable, and, Quinean doubts about the intelligibility of analyticity notwithstanding, the most satisfactory explanation would seem to be one where it is shown how the statement in question can be derived from logical truths in the narrow sense, simply by the substitution of definitionally equivalent expressions. The point, in short, is that acceptance of the view that the concept of temporal priority is analytically basic rules out the most satisfactory sort of explanation that might be offered of why it is a necessary truth, for example, that no event can be earlier than itself.

It seems plausible, then, to hold that the concept of temporal priority must be analysable. But how is it to be analysed? My own view, which I shall be touching upon in Chapter 9, is that the concept of temporal priority is to be analysed in causal terms. An advocate of the present approach to the direction of causation needs, however, to locate a plausible alternative. Can this be done?

[1] Suppes, *A Probabilistic Theory of Causation*, Amsterdam, 1970, p. 12.

I cannot survey all the possibilities here, but two alternatives that have been rather widely accepted may serve to illustrate the sort of problems that arise at this point for the attempt to identify the direction of causation with the direction of time.

One account of temporal priority accepted by a number of philosophers involves the view that the concept of temporal priority can be analysed in terms of the concepts of past, present, and future. In section 9.3, however, I shall indicate why it seems unlikely that such a claim can be sustained.

Another view which has been advanced fairly frequently is that the direction of time can be analysed in terms of the direction of irreversible processes. But if that sort of analysis is embraced, the present account of the direction of causation reduces to accounts which are, I shall argue in section 7.4, untenable.

What I am suggesting, in short, is that the analyses of the relation of temporal priority which are open to philosophers who wish to identify the direction of causation with the direction of time are either themselves exposed to decisive objections, or else such as, when combined with the present approach to the direction of causation, reduce the latter to other accounts which are demonstrably untenable.

The second and third objections are directed against the claim that one state of affairs can be causally prior to another only if it is also temporally prior. The thrust of the second objection is that it is logically possible for a cause to be simultaneous with its effect, while the third objection claims that it is even possible for an effect to precede its cause.

There are two main ways of attempting to support the claim that it is possible for cause and effect to be simultaneous. The one is to describe cases, either actual or possible, in which it seems plausible to say that a cause is simultaneous with its effect. The other involves supporting the claim that causes *can* be simultaneous with their effects by arguing for the stronger thesis that they *must* be.

First, the more modest claim that it is at least possible for cause and effect to be simultaneous. Can one point to actual cases in support of this claim? A typical case that might be offered is this. Consider a pencil that begins to move when a force is applied to one end of it. It is natural to suppose that the force that is applied to one end of the pencil causes the simultaneous motion of the other end. That this supposition cannot be correct is clear, however, in view of the fact that it is incompatible with the theory of relativity. But what exactly is wrong with the case? The answer is that, since our world does not contain any perfectly rigid bodies, it is not true that at the precise instant when the force is applied to the one end, the other end will immediately begin to move. What happens when the force is applied is that the end of the pencil to which it is applied is accelerated, and undergoes compression, with the result that the other end of the pencil does not begin to move until compression has occurred throughout the length of the pencil. So it is not the case that the force exerted at a given instant causally brings about, at the very same instant, the movement of the other end of the pencil.

The problem here is, of course, a very general one; it is not tied to the particular example just considered. For if relativity theory is correct, there is a finite limit to the speed with which causal processes can be transmitted, so actual examples of simultaneous, causally related events can never involve events that are spatially separated.

It might seem that this still leaves some room for actual examples of simultaneous, causally related states of affairs. What about cases of physical objects that are in contact? Will it not be true that, if one of the objects is moved, it is the movement of the surface of that object which is the cause of the movement of the part of the surface of the other object that is in contact with it? If so, the claim that this is a case where cause and effect are simultaneous will not conflict with relativity theory, since there is no distance between the two surfaces.

But this sort of example will not do either. The basic problem is this: on the one hand, if the relevant part of the surface of either object has any thickness, then the same sort of difficulty will arise as in the case of the pencil, while on the other, if neither surface has any thickness, they cannot be in contact. For if the surfaces have no thickness, they can be represented by closed sets of points, and two such sets cannot stand in the relation that is required if the corresponding objects are to be in contact.

As a consequence it seems likely that our world does not, in fact, contain any cases in which a cause and its effect are simultaneous. It seems, therefore, that one must instead consider whether one can describe logically possible cases in which a cause and its effect would be simultaneous.

A natural line of thought is this: imagine a world rather like ours, but where relativity theory is false; in such a world light might travel with infinite velocity, or there might be perfectly rigid bodies. If the former were the case, then the propagation of a ray of light would be a causal process that did not take any time. So earlier stages in the propagation of the wave would give rise to later stages which were simultaneous with the earlier ones. Alternatively, if the world were one that contained perfectly rigid bodies, then one could appeal to the case of the pencil. For then there would be no interval between the time at which the force was applied to one end of the perfectly rigid pencil and the time at which the other end began to accelerate. So here, too, one would have a case where cause and effect were simultaneous.

Are such hypothetical cases convincing? Do they show that it is logically possible for a cause and its effect to be simultaneous? Perhaps—though it seems to me that the issue is by no means clear-cut. Consider a world where light travels with infinite velocity. If a light source is turned on, and light travels from it to some object, is it really satisfactory to say that the event which is the light's leaving the source is simultaneous with the event which is the light's striking the object? Might not one instead say that, while it is true that if standard real numbers are used to represent moments in time, the same number must be assigned to both events, with the result that there is no temporal gap between them, what this sort of case really shows is that in a world where light travelled with infinite velocity, moments in time could not be adequately represented by means of the standard real numbers? Just as, in the case of probability, one has to introduce infinitesimals if one is to have a satisfactory representation of the probabilities of certain occurrences, so, in a world where some things travel with unlimited velocities, infinitesimals must be introduced if one is to have an adequate representation of time. And when this is done, it will not be true that the light strikes the object at the same time that it leaves the source: there will be an infinitesimal time interval between the two events.

Is this view of the matter correct? I am inclined to think that it is, though I do not see how the issue can be argued other than by appealing to a causal theory of time.

Let us now turn to the second, and rather more dramatic, way of attempting to establish the claim that causes can be simultaneous with their effects, namely, by arguing that they *must* be. On the face of it, this line of thought seems rather unpromising. For as Richard Taylor has argued, if all causes were contemporaneous with their effects, temporally separated states of affairs could never be causally related.[2] And surely if we know anything at all about the causal relations that there are in the world, we know that temporally separated events are sometimes causally related!

At least one philosopher, however, has argued for the view that it is a mistake to think that temporally separated states of affairs are ever causally related: Myles Brand, in his article 'Simultaneous Causation'.[3] The argument that Brand offers rests upon what he refers to as 'Hume's maxim':

> For any events *e* and *f* and time interval *t*, if *e* occurs during *t* and *e* does not change during *t* and *e* is the cause of *f*, then *f* occurs during *t*.[4]

But what reason is there for accepting this claim? Brand's response is as follows:

Hume's maxim says that if an event *e* occurs during *t* and does not *genuinely* change during *t*, then *f* also occurs during *t* if *e* is the cause of *f*. Suppose that, contrary to the maxim, *e* occurred during *t*, *e* remained the same during *t*, *e* was the cause of *f*, but *f* began after *e* began. However, if *f* began after *e* began, then something happened after *e* began to have made *f* begin. There was nothing, however, that happened after *e* began that could have made *f* begin. No other event could have made *f* begin after *e* began, since *e* was the cause of *f*; and no change in *e* during *t* could have made *f* begin after *e* started, since *e* did not change during *t*. Thus, *f* did not begin after *e* began, and the denial of the maxim does not specify a situation that can occur.[5]

This argument seems weak. In the first place, what is logically untoward about there being a causal law expressed by a statement of the form 'If *x* has property *P* at time *t*, that state of affairs will directly give rise to the existence of some other entity, *y*, that will have property *Q* at time *t**', where there is a non-infinitesimal time

[2] Taylor, *Action and Purpose*, Englewood Cliffs, NJ, 1966, p. 38.
[3] Brand, 'Simultaneous Causation', *Time and Cause*, ed. van Inwagen, Dordrecht, 1980, pp. 137–53.
[4] Ibid., p. 147. [5] Ibid., p. 148.

interval, d, between t and t^*? How does Brand, in the above argument, show that there cannot be such gappy causal connections? The answer is that he has not really offered any argument. He has simply *asserted* that '. . . if f began after e began, then something happened after e began to have made f begin'.

Secondly, the situation is the same if one sets aside the possibility of gappy causal laws, and considers the less esoteric case of laws according to which the value of some quantity at some point in time is causally dependent upon the value of certain quantities during a preceding interval. Thus, consider a law which says that the velocity a particle has at some time t_1 is causally determined by its velocity at some earlier time t_0 together with all of the forces that acted upon the particle during the interval from t_0 up to, *but not including*, time t_1. Now if there were no forces acting upon the particle during that time interval, one would have a case that falls under Hume's maxim, since the velocity of the particle does not change during the interval preceding time t_1, but it would be a case that falsifies the maxim, since the effect does not occur at the same time as the cause. Brand needs to show that this sort of case is logically impossible. But here too, the only thing in his argument that is relevant to this case is the unsupported assertion that if f began after e began, then something happened after e began to have made f begin.

These cases show, moreover, that the problem is not merely with Brand's argument. Since both sorts of case are logically incompatible with Hume's maxim, these possibilities show that that maxim cannot express a logically necessary truth.

The final question that needs to be considered is whether it would follow, if Hume's maxim were a necessary truth, that a cause must be simultaneous with its effect. Brand does not develop his argument here very carefully, but his basic line of thought appears to be this. Causes must be either events that involve change or events that do not. If the cause is an unchanging event, then Hume's maxim leads immediately to the conclusion that the effect must be simultaneous with the cause. Suppose, then, that the event involves change. It can then be resolved into a sequence of unchanging events, the final member of which will be the cause of the original effect. Hume's maxim can then be seen to hold.[6]

[6] Ibid., p. 148.

The main problem with this argument lies in the assumption that if one can cite some changing event as a cause of some other event, the former event can be resolved into a sequence of unchanging events having a final member which is the cause of the original effect. That this is false can be seen from the following sort of case. A particle has velocity v_0 at time t_0, and during the interval from time t_0 up to, but not including, time t_1, is acted upon by a force that varies continuously. The particle's having velocity v_0 at time t_0, together with the forces acting upon the particle during the interval, are causally sufficient to ensure that the particle has velocity v_1 at time t_1. Can this cause be resolved into a sequence of unchanging events, the final member of which can be cited as the cause of the particle's having velocity v_1 at time t_1? The answer, obviously, is that it cannot. For no matter how small an interval, t, prior to time t_1, is selected, the velocity of the particle, and the forces acting upon it, are varying throughout that interval. There does not exist, therefore, any *unchanging* event which is the cause of the particle's having velocity v_1 at time t_1. So Hume's maxim cannot be applied.

Brand's overall argument is, in short, rather unsatisfactory. His defence of Hume's maxim, upon which the argument turns, is unsuccessful. Moreover, the maxim itself does not appear to express a logically necessary truth. Finally, even if it did, it would not follow that causes must be simultaneous with their effects.

Let us now turn to the third objection to the view that the asymmetry and direction of causal relations are to be explained in terms of the corresponding properties of the relation of temporal priority, namely, the contention that it is logically possible for an effect to precede its cause. Given the difficulty of establishing even the more moderate claim that it is possible for an effect to be simultaneous with its cause, this third line of attack may seem very unpromising. I shall attempt to show, however, that although the claim cannot be sustained as it stands, it is possible to set out a closely related contention that does appear to be correct.

The question of whether it is possible for an effect to precede its cause has been discussed very extensively, and most philosophers seem to feel that a negative answer is called for.[7] A variety of

---

[7] Sustained discussion of this topic was initiated by the symposium involving Dummett and Flew, 'Can an Effect Precede its Cause?', *Proceedings of the Aristotelian Society*, Suppl. 28, 1954, pp. 27–62. Further contributions include:

grounds has been offered in support of this view of the matter. Some have argued that allowing both forward and backward causation gives rise to contradictions. Others have contended that no claim that later states of affairs of one type are causally both necessary and sufficient for earlier states of some other type could possibly be true, since one could always make such statements false by waiting until the earlier states occurred, and then arranging things so that the supposed cause could not occur. However, the objection that probably has the strongest intuitive appeal centres around the claim that the past is fixed, or settled, or already in existence while the future is not, together with the claim that what is fixed or settled, or already in existence, cannot be caused by what is not fixed or settled, or by what does not yet exist.

What I wish to do is to consider what seems to be the strongest argument in support of the claim that it is logically possible for an effect to precede its cause. The argument I have in mind was advanced by Michael Dummett in his article, 'Bringing about the Past'.[8] The force of Dummett's argument has not always been fully appreciated however, so it may be useful to offer a somewhat modified version of it. We shall see that it is possible to resist the argument, but I think that this can only be done if one holds that the concept of temporal priority presupposes the concept of causal priority, rather than vice versa.

The modified argument may be put as follows. Imagine a universe that initially seems very like our own, but which turns out to differ from it in certain rather striking ways. A wall is discovered with some quite remarkable properties. The wall is indestructible, and there is no way of getting round it. Neither matter nor energy can pass from our side to the other. However there are areas of the wall that function as one-way windows transmitting both light waves and sound waves from the other side,

Black, 'Why Cannot an Effect Precede its Cause?', *Analysis*, 16, 1955–6, pp. 49–58; Flew, 'Effects before their Causes—Addenda and Corrigenda', ibid., pp. 104–10; Scriven, 'Randomness and the Causal Order', *Analysis*, 17, 1956–7, pp. 5–9; Pears, 'The Priority of Causes', ibid., pp. 54–63; Flew, 'Causal Disorder Again', ibid., pp. 81–6; Dummett, 'Bringing About the Past', *Philosophical Review*, 73, 1964, pp. 338–59; Gorovitz, 'Leaving the Past Alone', ibid., pp. 360–71; Gale, 'Why a Cause Cannot be Later than its Effect', *Review of Metaphysics*, 19, 1965, pp. 209–34.

[8] Dummett, op. cit., pp. 339–40. See also the discussion by Reichenbach, in *The Direction of Time*, Berkeley, 1956, pp. 139–40.

thus making it possible for us to observe what is transpiring there. It turns out that the world on the far side is very similar to our own, with one small exception: all processes appear to be happening in reverse. For example, we see a piece of paper on a desk, the ink on it completely dry. The ink slowly becomes moist, and a man walks backwards into the room, pen in hand, and sits down at the desk. He begins, so it seems, to retrace the writing on the paper, starting on the bottom line and moving from right to left. As he does so, we notice that not only does he retrace the writing perfectly, but the ink moves up into the pen, so that he is left with a blank piece of paper. We then hear some sound, initially unintelligible, but which when recorded and played back in reverse, becomes 'I'll write this letter before I go golfing.'

How should one describe such a world, in causal and temporal terms? First, the causal description. If it were a matter of single, much more limited occurrence, it would probably be reasonable to view the sequence of events as an accident merely having the appearance of a causally reversed process. But an occurrence of the sort described would require a very large number of remarkable coincidences if the direction of causation were the same as in our part of the world, or if, still more remarkably, there were no causal processes at all on the other side. And if one supposes that *all* processes on the other side appear reversed, and that they are at least as common as the non-reversed ones on this side, then both the hypothesis that the causal direction is the same on both sides, and the hypothesis that there are no causal processes at all over there, surely become fantastically improbable. The reasonable conclusion, then, would seem to be that causal processes run in opposite directions on different sides of the wall.

It is, however, possible to dispute this. Causal relations have a number of structural properties. It is, for example, conceptually impossible for two events to cause each other. But, given that this is so, may it not also be the case that our concept of causation is such that the hypothesis that there are causal processes running in opposite directions is *incoherent*? If so, there will be no alternative to saying either that the direction of causal processes is the same on both sides, or that there are no causal processes at all on at least one side of the wall.

It is, perhaps, unclear whether our ordinary causal concepts have the relevant structural property. My own view is that they do not. But even if they do, I believe that the basic argument remains intact. The reason is this: given a relation that necessarily has certain structural properties, it is usually possible to define it in terms of some other relations which need not have those properties. Consider, for example, the relation of being heavier than. This relation is necessarily transitive. But it would seem that there is an underlying relation, which is not necessarily transitive, and in terms of which the relation of being heavier than can be defined. For let us say that *A* outbalances *B* if and only if whenever *A* and *B* are put on opposite sides of a standard scale, the side with *A* on it moves downward. One can easily imagine what it would be like for *A* to outbalance *B*, *B* to outbalance *C*, and yet for *C* to outbalance *A*. So it is not logically necessary that the relation of outbalancing be transitive. None the less, it is easy to define the necessarily transitive relation of being heavier than in terms of the relation of outbalancing: '*A* is heavier than *B*' means the same as '*A* outbalances *B*, and it is a law that, for all *X*, *Y* and *Z*, if *X* outbalances *Y*, and *Y* outbalances *Z*, then *X* outbalances *Z*.'

The idea, then, is to argue that even if causal relations, as ordinarily conceived, necessarily possess the structural property in question, so that it is conceptually impossible for causal processes to run in opposite directions, there is some weaker, underlying relation—call it proto-causation—which need not have that structural property, but which does have a direction associated with it, and which holds between any two states of affairs that are causally related. The contention will then be first, that the argument concerning the unusual world can now be taken as showing that the reasonable hypothesis is that proto-causation runs in opposite directions on different sides of the wall, and secondly, that since the relation of proto-causation is analytically more basic than causation as ordinarily conceived, it is plausible to hold that the direction of causation is to be understood in terms of the direction of the underlying relation of proto-causation. If so, the question becomes one of the relation between proto-causal priority, and temporal priority.

The next question concerns the temporal description of our two-part world. One alternative is to say that the direction of time is the same on both sides of the wall. If this alternative is adopted it follows, if the preceding argument is correct, that there can be cases in which causes, or at least proto-causes, come after their effects.

The alternative is to abandon the notion of a universal temporal order, and to relativize temporal concepts to different regions. There is, however, some reason for thinking that this cannot always be done without central features of our ordinary conception of time being lost. Consider, for example, the distinction between past and future. Many would probably maintain that it is essential to this distinction that the past is fixed and settled in some sense in which the future is not. But what are we to say about events on the other side of the wall which are, so to speak, future events relative to an appropriate coordinate system within that region? The problem is that they are events that *we* may have already observed, and hence ones that are already settled.

Let us suppose, however, that we have a concept of temporal priority which does not involve any distinction between events that are fixed and settled, and those that are not, and that it is this concept of temporal priority that we are relativizing to different regions. If we do this, we shall be able to avoid accepting the view that, in the world envisaged, some effects precede their causes, or proto-causes. However this way of saving the thesis than an effect cannot precede its cause appears to preclude the possibility of analysing causal priority in terms of temporal priority, since when temporal priority is thus relativized to a region, the direction that is assigned to time in a given region seems to be based upon the direction of causal, or proto-causal processes, within that region.

The upshot is that the above argument seems to support the conclusion that *either* it is logically possible for an effect to precede its cause, or proto-cause, *or else* causal, or proto-causal concepts are more basic than temporal, so that the former cannot be analysed in terms of the latter. In short, the third objection cannot be answered without granting the first. We would seem to be justified in concluding, then, that the asymmetry of causal processes, and the direction of them, cannot be explained in terms of the corresponding properties of the relation of temporal priority.

7.2 MACKIE'S ACCOUNT OF CAUSAL PRIORITY

A second and related attempt to provide an account of causal priority is that advanced by John Mackie:

Suppose that X and Y are individual events, and X is seen as necessary (and sufficient) in the circumstances for Y, so that the basic requirement for the judgement that X caused Y is met. Then, despite this, X was not causally prior to Y if there was a time at which Y was fixed while X was unfixed. If, on the other hand, X was fixed at a time when Y was unfixed, then X was causally prior to Y. Again, if X was not fixed until it occurred, then even if Y also was fixed as soon as X occurred (given, of course, that X was necessary in the circumstances for Y), X was causally prior to Y. And further, if there is some line or chain of causation, some continuous causal process, linking X and Y and some other event Z so that X was between Y and Z, and if Z was not fixed until it occurred, then X was causally prior to Y.[9]

What does it mean to say that an event is *fixed* at a given time? Mackie's construal appears to be that an event e occurring at time t, is fixed at time t^* if and only if either t^* is not earlier than t, or there is some event d which is nomologically sufficient for e, and which occurs at or before time t^*.

Mackie's analysis of causal priority appears to be open to a number of objections. First, as he himself points out, his account implies that in a totally deterministic world—understood here as a world in which, for every contingent state of affairs involving particulars, there is a temporally prior, nomologically sufficient condition—no states of affairs would stand in the relation of causal priority: '. . . if *total* determinism holds, and there was not even a first creative event, our present concept of causal priority will not be true of the real world'.[10]

For on Mackie's account, relations of causal priority can obtain between events only if there is some time at which the events are not fixed, and in a totally deterministic universe this will not be so. Mackie was not especially troubled by this consequence. Yet surely it is rather surprising if our ordinary concept of causation has the consequence that it is logically necessary that some events be causally undetermined.

Secondly, in analysing causal priority in terms of fixity, which in turn involves the notion of temporal priority, Mackie faces the problem that confronts any attempt to relate causal priority to temporal priority, namely, what account can be given of temporal priority? Can it plausibly be treated as primitive, or is some

[9] Mackie, *The Cement of the Universe*, Oxford, 1974, p. 190.
[10] Ibid., pp. 191–2.

analysis called for? And if the concept of temporal priority stands in need of analysis, is there any satisfactory alternative that does not involve causal notions? If not, then Mackie's account is implicitly circular.

A third objection arises from a possibility mentioned by Douglas Gasking.[11] Suppose that iron glows in a certain way only when its temperature is at least 1000 °C. It seems possible that the state of affairs which is the iron's being at that temperature might have precisely the same spatiotemporal location as the state of affairs which is the iron's glowing. But even if that were so, one could, Gasking argues, know that the former was causally prior to the latter, in view of the fact that the heating of substances always makes them hotter, but does not in general make them glow in the way that iron glows at 1000 °C.

But why does this sort of case constitute a problem for Mackie's analysis? The reason is this: first, given that the events occur at the same time, and that each is nomologically sufficient for the other, there cannot be any time at which one is fixed, and the other not. And second, given that they also occur at the same place, there can be no intervening causal process by virtue of which one is causally prior to the other. So if such cases do occur, or if, at least, they are at least logically possible, Mackie's analysis must be unsound.

Whether this objection has force depends upon whether it is logically possible for a cause and its effect to be simultaneous. In the previous section, I argued that the idea that it is possible for an effect to be simultaneous with its cause is more problematic than it at first appears. If, as I am inclined to think, that idea must in fact be rejected, the present objection will have to be set aside.

A final objection is that Mackie's analysis, like the preceding account of causal priority in terms of temporal priority, has difficulty with the possibility of a world in which causal, or at least proto-causal processes, run in opposite directions. Thus, on the one hand, if one postulates a single direction for time which determines what events are fixed at a given time on both sides of the wall, it will follow that there are cases in which event *e* is fixed at time *t*, and *c* not, even though *e* is causally, or proto-causally, dependent upon *c*. On the other hand, if the direction of time is

---

[11] Gasking, 'Causation and Recipes', *Mind*, 64, 1955, pp. 479–87.

different on opposite sides of the wall, it seems that the direction of temporal priority, and with it what events are fixed at a given time, must rest upon causal or proto-causal relations. For otherwise, what account can possibly be given of the grounds for holding that time has one direction, rather than another, in a given region? The possibility of a world in which causal or proto-causal processes run in different directions thus seems to show that Mackie's analysis must be either false or circular.

7.3 REICHENBACH'S MARK METHOD

In his book, *The Philosophy of Space and Time*, Hans Reichenbach proposed what has been called the *mark method* for defining the direction of causal processes:

Causality establishes not a symmetrical but an asymmetrical relation between two events. If we represent the cause–effect relation by the symbol C, the two cases

$C(E_1, E_2)$ and $C(E_2, E_1)$

can be distinguished; experience tells us which of the two cases actually occurs. We can state this distinction as follows:

If E_1 is the cause of E_2, then a small variation (a mark) *in E_1 is associated with a small variation in E_2, whereas small variations in E_2 are not associated with variations in E_1.*

If we wish to express even more clearly that this formulation does not contain the concept of temporal order, we can express it in the following form, where the events that show a slight variation are designated by E^*:

We observe only the combinations

$E_1E_2 \quad E_1{}^*E_2{}^* \quad E_1E_2{}^*$

and never the combination

$E_1{}^*E_2.$ [12]

Thus, to use Reichenbach's own illustrations, if a beam of white light is travelling from A to B, and a red filter is inserted at point A, the light will be red both at point A and at point B, whereas if

---

[12] Reichenbach, *The Philosophy of Space and Time*, New York, 1958, sect. 21, pp. 136–7. This is a translation of his Philosophie der Raum-Zeit-Lehre, Berlin, 1928.

the filter is inserted at B, the light will be red at B but not at A. Or if a stone is travelling from A to B, and a chalk mark is placed on it at A, it will also have a chalk mark at B, whereas when a chalk mark is placed on it at point B, none is found at point A.

This approach is exposed to a number of objections. In the first place, consider a causal process that consists of a sequence of events of the same sort—such as the movement of a billiard ball with uniform velocity. On Reichenbach's account, the direction of a causal processes is determined by the fact that the combinations E_1E_2, $E_1^*E_2^*$, and $E_1E_2^*$ do occur, while the combination $E_1^*E_2$ does not. In the case of a causal sequence of events of the same sort, the direction of the process would then be determined by the fact that the combinations E_1E_1, $E_1^*E_1^*$, and $E_1E_1^*$ do occur, while the combination $E_1^*E_1$ does not. But this is incoherent, since given that the notation E_1E_2 does not, as Reichenbach emphasizes, incorporate any reference to temporal order, the combination $E_1E_1^*$ is identical with the combination $E_1^*E_1$. Consequently, Reichenbach's approach is inapplicable to the case of causal processes that consist of sequences of events of the same sort.

A second objection is this: consider a very simple universe, such as one involving only neutrons, forever rotating about one another. Intuitively, one certainly wants to speak of causal relations in such a simple universe: the one neutron moves in a circular orbit due to the gravitational field that is present by virtue of the other neutron. But on Reichenbach's account, such a world would be devoid of causation, since nothing ever gets marked at any time.

A third objection to Reichenbach's approach to the direction of causal processes is that the mark method cannot lead to satisfactory results unless the methods of marking are restricted to those that involve *irreversible* processes.[13] For consider the case where a man is walking along, wearing a hat, which is blown away by the wind. Until the direction of causation has been specified this could equally well me a case where a mark, in the form of a hat, is imposed by the wind upon a man who is walking backwards.

The problem in general is that what, viewed from one direction,

[13] Compare Grünbaum's discussion in 'Carnap's Views on the Foundations of Geometry', *The Philosophy of Rudolf Carnap*, ed. Schilpp, La Salle, Ill. 1963, pp. 599–684. See pp. 605–7.

is a case of imposing a mark is, when viewed from the other direction, a case of removing a mark. So it would seem that in classifying a certain event as a case of marking, rather than as one of removing a mark previously imposed, one is tacitly assuming that the direction of time is given.

This objection will not arise if one alters Reichenbach's method so that one considers only those methods of applying marks which are such that they cannot be removed via the reverse process. But when the mark method is thus restricted, one no longer has an independent way of establishing the direction of causal processes. It becomes just a special case of the more general attempt to establish a direction for causation by appealing to irreversible processes—as was recognized by Reichenbach in his later work.[14]

7.4 ENTROPIC AND NON-ENTROPIC IRREVERSIBILITY

The most important alternatives to a causal account of the direction of time are those that appeal to the direction of irreversible processes. Such approaches are of two main types. First, there are those that centre upon the concept of entropy, and which, as a first approximation, equate the direction of time with the direction of increase of entropy in closed systems. Second, there are those that relate the direction of time to the direction of processes that exhibit non-entropic irreversibility. In this section I wish to consider whether either of those approaches can provide an account of causal priority.

First, then, the view that the direction of time is the direction of increase in entropy in closed systems. The reason that this is only a first approximation is as follows. The so-called 'phenomenological' version of the second law of thermodynamics, which originated with Clausius, asserts that entropy *never* decreases in any closed system. If this were a law, it might be possible to identify the direction of time with the direction of increase in entropy in all closed systems. But the problem is that there is excellent reason for holding that the phenomenological version of the second law of thermodynamics is simply false. For when one considers the laws governing the micro-states that ultimately determine the level of

---

[14] Reichenbach, *The Direction of Time*, sect. 23, pp. 197–8.

entropy in a system, it becomes clear that, not merely is it nomologically possible for the entropy of a closed system to decrease, but that there is, in general, some non-zero, non-infinitesimal probability that a given closed system will, during a given interval, undergo a decrease in entropy of a certain magnitude.

Suppose, then, that the phenomenological version of the second law of thermodynamics is set aside, and one turns instead to the statistical version formulated by Boltzmann and Gibbs.[15] The natural move is then to equate the direction of time with the direction of increase in entropy in *the majority* of closed systems. But this idea, unfortunately, would involve a misunderstanding regarding the content of the statistical version of the second law of thermodynamics. The law does not assert that entropy is likely to increase in a closed system. It asserts rather that if one has a closed system that is in a state of low entropy, then the probability that its entropy will increase is very high. But this latter assertion is perfectly compatible with the assertion that if one has a closed system in a state of low entropy, then the probability is also very high that it was previously in a state of higher entropy. Indeed, the method employed to establish that a closed system with low entropy is very likely to change into a state of higher entropy can be used to show that it is *just as likely* that such a system developed out of a system that enjoyed higher entropy.[16] This in turn means that the statistical version of the second law of thermodynamics does *not* entail that entropy is likely to increase in a closed system: decreases in entropy are precisely as likely as increases.

The situation, in short, is this: on the one hand, the phenomenological version of the second law of thermodynamics is asymmetric with respect to time, so if it were correct, one might be able to appeal to it to establish a direction for time. But the phenomeno-

---

[15] Boltzmann, Vorlesung über die kinetische Theorie der Gase, Leipzig, 1902. Gibbs, *Elementary Principles in Statistical Mechanics*, New York, 1922.

[16] These results were established by P. and T. Ehrenfest, 'Begriffliche Grundlagen der statistischen Auffassung in der Mechanik', Encyclopädie der mathematischen Wissenschaften, 4/2, II, Leipzig, 1911, pp. 41–51, and by Smoluchowski, 'Gültigkeitsgrenzen des Zweiten Hauptsatzes der Warmetheorie', *Œuvres*, 2, 1927, pp. 361–98. For further discussion, see Mehlberg, 'Physical Laws and Time's Arrow', *Current Issues in the Philosophy of Science*, ed. Feigl and Maxwell, New York, 1961, esp. pp. 113–16; Reichenbach, op. cit., pp. 108–17; Grünbaum, op. cit., pp. 640–4, and chap. 8, *Philosophical Problems of Space and Time*, 2nd edn, Dordrecht, 1973, esp. pp. 236–42.

logical version does not express a law. The statistical version of the second law, in contrast, is defensible. But it is not asymmetric with respect to time. It appears, then, that there is no prospect of defining a direction for time by reference to the laws of thermodynamics.

Philosophers such as Reichenbach and Grünbaum have argued, however, that it may still be possible to define the direction of time by reference, not to the laws of thermodynamics, but to certain striking, non-nomological facts about irreversible thermodynamic processes as actually found in the region of the universe that we inhabit. Thus Reichenbach introduces what he calls the hypothesis of the branch structure,[17] which may be explained as follows. A *branch system* is part of a more comprehensive system which breaks off from it, and then, while remaining relatively isolated from the larger system for a period of time, develops from a state of relatively low entropy to one of higher entropy:

> A rock embedded in snow is heated by the sun; during this process, the system consisting of the sun and the rock increases its entropy because of the absorption of radiation by the rock. At night, the rock and the snow together form a system which, because of its inner temperature differences, has a low entropy; and the cooling of the rock and the melting of the snow represent a transition to higher entropy.[18]

Reichenbach says that the world around us abounds in such branch systems, and this observation leads him to formulate his hypothesis of the branch structure. The central claim advanced by this hypothesis is that the universe, at present, involves a large number of branch systems, and that the direction of change of entropy is the same, both in the vast majority of such branch systems, and in the more comprehensive systems to which they originally belonged, and which they in time rejoin.[19]

How does the introduction of reference to branch systems help one to deal with the problem of defining a direction for time? The answer is that the behaviour of branch systems differs in one very important respect from the behaviour of permanently closed systems. In the case of permanently closed systems, it is certainly true that the vast majority that are in a state of low entropy at a given time will be in a state of higher entropy at a later time. But

---

[17] Reichenbach, op. cit., sect. 14–16, pp. 117–43.
[18] Ibid., p. 118.   [19] Ibid., p. 136.

they will *also* have been in a state of higher entropy at an *earlier* time. By contrast, in the case of branch systems that are in a state of low entropy, the vast majority will also be in a higher-entropy state at later times, but they will *not* have been in a higher-entropy state at earlier times, since they did not then exist as separate systems. So by switching from consideration of permanently closed systems to consideration of branch systems, one can—if one accepts the hypothesis of the branch structure—identify the positive direction of time with the direction of increase in entropy in the majority of branch systems with low entropy.[20]

The other main non-causal approach to the problem of the direction of time involves non-entropic irreversibility. This approach, advocated by Karl Popper, centres around the notion of *dispersal of order*.[21] Popper mentions the case of outgoing circular waves produced when an object strikes the surface of a pond. Such sequences are commonplace, while the reverse sequences, though certainly compatible with the laws of nature, never occur spontaneously.

Now it is true that such processes generally involve an increase in entropy. However Popper's point is that there is a non-entropic feature of such processes that makes them *de facto* irreversible: the dispersal of order from a central source. For although it is conceivable that the reverse process might occur spontaneously, the likelihood of that happening is virtually zero. If such a process involving converging circular waves were to be found, it would almost certainly be due to a coordination of generators at the periphery from which the incoming waves originate, and so would not be spontaneous. Popper's suggestion, then, is that the direction of irreversible processes involving dispersal of order can be used to explain the anisotropy of time.

Can either of these approaches to the problem of the direction of time be used to provide an account of causal priority? I believe that there are at least three reasons for thinking that they cannot. In the first place, consider some everyday causal claims: (1) the glass was accelerated by the force of gravity that the earth exerted upon it; (2) the rapid acceleration of the golf ball was due to the

---

[20] For further discussion, see Grünbaum, op. cit., pp. 254–64.
[21] Popper, 'The Arrow of Time', *Nature*, 177, 1956, p. 538. There is further discussion in three subsequent notes by Popper which appeared in *Nature*: 178, 1956, p. 382; 179, 1957, p. 1297; 181, 1958, pp. 402–3.

impact of the club. In the first of these cases there is neither dispersal of order nor necessarily a change in entropy. In the second case there is an increase in entropy, but one's belief that there is a causal connection between the acceleration of the ball and the impact of the club is surely completely independent of any belief that there is an accompanying increase in entropy due, for example, to the conversion of some kinetic energy into heat. So it would seem that many of the most familiar causal connections are not such that the causal priority involved is to be explained in terms either of the direction of dispersal of order or of the direction of increase in entropy.

This first objection assumes, however, that the analysis being offered has the following form:

> State of affairs *a* is causally prior to state of affairs *b* if and only if *a* and *b* are part of a process that involves dispersal of order (or increase in entropy) in the direction from *a* to *b*.

In reply, a slightly more complicated account might be suggested, according to which whether state of affairs *a* is causally prior to state of affairs *b* is not always determined by features of *a*, *b*, and the process connecting them. Reichenbach, for example, suggested that one might appeal to the reversible laws of mechanics to establish a causal net connecting events. Initially, of course, there would be no direction associated with any of the connections. Thus, if *a* was causally connected with both *b* and *c*, one would not know whether (1) *a* was causally prior to both *b* and *c*, or (2) *b* and *c* were both causally prior to *a*, or (3) *b* was causally prior to *a*, which was prior to *c*, or (4) *c* was causally prior to *a*, which was causally prior to *b*. Reichenbach went on to argue, however, that by appealing to what he referred to as 'the principle of the local comparability of the time order', one could determine whether the two causal links—the one involving *b* and *a*, the other involving *a* and *c*—were equidirected or counterdirected. This would allow one to establish an overall order for the net, though not a direction. One could then go on to consider irreversible processes. Not all events, of course, would be involved in such processes. However one could appeal to the direction of irreversible processes in the cases of events that were so involved, and the result would serve to establish a direction for the net as a whole, including those pairs of events not involved in irreversible processes.[22]

The fundamental idea, therefore, is that relations of causal priority are, so to speak, defined directly for some pairs of states of affairs, and then transferred to other pairs for which they cannot be so defined. But now it can be objected that this idea is strongly counter-intuitive. For it implies, in the first place, that whether *a* is the cause of *b* may depend upon facts about causal processes other than the one connecting *a* and *b*, and in the second place, that whether *a* is the cause of *b* is something that may not be ontologically fixed until some time long after the occurrence of both *a* and *b*. These consequences are, I suggest, quite unacceptable. Whether *a* is the cause of *b* does not depend upon other causal processes, let alone upon states of affairs existing long after the occurrence of *a* and *b*.

A second, and perhaps even more serious objection to these attempts to analyse causal priority in terms of irreversible processes is this. Suppose that the world is slightly different than we believe it to be, and that every billion years the universe is subject to a massive, random perturbation. Suppose further that our world is a completely deterministic one, and that its laws are invariant with respect to a reversal of the temporal axis. Finally, assume that the next shake-up will take place in the middle of the year 2000. Among the possible, albeit highly unlikely outcomes of such a grand shake-up, is one where everything is unchanged, except for the fact that all velocities have been reversed. Then, given that we are assuming that the laws are completely deterministic, and invariant with respect to time reversal, the world will pass through states exactly like those it passed through before, except that all velocities will be reversed. Since entropy will, in our world, presumably increase during the period from 1990 to the middle of the year 2000 in the vast majority of relevant systems, it will have to decrease in the vast majority of those systems from the middle of 2000 to 2010. The direction of increase in entropy will therefore run from 2010 back to the middle of the year 2000. Similarly, just as order will appear to be dispersed in various processes in the direction from 1990 to the middle of 2000, so it must be the case that it will appear that order is being dispersed, in

_____

[22] Reichenbach, op. cit., pp. 27–36.

precisely the same sorts of processes, in the direction from 2010 back to the middle of 2000. But by hypothesis, the direction of causation runs from events in the middle of 2000 to events in 2010 and beyond. So in the world envisaged, the direction of causal processes will be, from the middle of 2000 on, contrary both to the direction of increase in entropy, and to the direction in which order is apparently being dispersed.

This argument could, of course, equally well be formulated in terms of two completely separate spatiotemporal worlds, similar to those appealed to in the 'inverted universes' argument in the previous chapter. A Laplacean deity works out the position and velocity of every particle in our world in the middle of 2000, and then creates another world just like it except for a reversal of all velocities. Ten years later—assuming that the laws are deterministic and invariant with respect to time reversal—the state of the new world will be the inverse of the state of our own world in 1990. So causal processes in the newly created world are running in a direction that is opposite both to the direction of increase in entropy, and to the direction of apparent dispersal of order.

It might be argued, however, that the fact that the second world has been created by making use of information that will ensure that it has the appearance of being inverted raises a problem. Is it not like the case where one has converging circular waves due to the action of coordinated generators at the edge of a pond? This objection is perhaps correct. But if so, one can easily reformulate the argument to avoid it. Rather than assuming that the new world is created by a Laplacean deity, one can suppose that the two inverted universes came about by a grand accident.

A third objection can be stated very briefly: it is the 'simple world' objection. Thus, consider the world that contains only two neutrons, endlessly rotating around one another. Surely it is true both that such a world is logically possible, and that it contains states of affairs that are causally related. But there is never any change in entropy, nor is there ever any dispersal of order. Therefore the direction of causation cannot be analysed either in terms of increase in entropy, or in terms of dispersal of order.

The conclusion, in short, is this. It may be true in our world that the direction of causal processes *coincides* with the direction of increase in entropy in the vast majority of branch systems, and with the direction in which order is transmitted in non-entropically irreversible processes. But the above three arguments show that causal priority cannot be *analysed* in terms of either of these types of irreversibility. First, because there are causal processes that do not exhibit irreversibility, and it is very implausible to hold that the direction of such processes is parasitic upon the direction of other causal processes. Second, because there are possible worlds in which the direction of causal processes does not coincide with the direction of irreversible processes. And third, because there are possible worlds that contain causal processes, but which are devoid of all irreversible processes, both entropic and non-entropic.

7.5 PROBABILISTIC RELATIONS AND CAUSAL PRIORITY

Hans Reichenbach advanced a number of suggestions concerning how one might make sense of the notion of the direction of causal processes. I have already discussed two of his proposals: one based upon the mark method, another involving an appeal to entropic behaviour exhibited by branch systems. A third important suggestion which he made was that all causal concepts, including that of the direction of causal processes, might be explained in terms of certain statistical relations.

This general approach has recently been carefully developed by Wesley Salmon in a number of interesting articles. Salmon does not, however, accept Reichenbach's claim that *all* causal concepts can be explicated entirely in terms of appropriate statistical relations among the types of event in question. He holds, as we shall see, that one needs to supplement statistical considerations with the notion of a causal process.

As the accounts of Reichenbach and Salmon do not appear to differ with respect to the concept that concerns us here—that of causal priority—and as Salmon's account seems preferable to Reichenbach's where they do diverge, my discussion will focus upon the account offered by Salmon.

Three ideas feature in Salmon's account of causal relations: (1) the idea of *positive statistical relevance*; (2) the idea of *causal processes*; (3) the idea of *causal interactions*. As regards the first, Salmon holds—and here he agrees with Reichenbach and

Suppes[23]—that a necessary condition for states of affairs of type B being causes of states of affairs of type A is that states of type B must stand in a relation of positive statistical relevance to states of affairs of type A, where the latter is the case if and only if

Prob(A, B) > Prob(A).

Secondly, the notion of a causal process: causal processes are to be contrasted with causal interactions. The latter are, very roughly, *loci* of change, whereas the former can be thought of as continuous series of states of affairs that share some common property, either qualitative or structural. Paradigm cases of causal processes are the movement of a material particle, and the propagation of a ray of light.

The main problem, as Salmon sees it, in offering a character-ization of causal processes, involves distinguishing between genuine causal processes and pseudo-processes. Consider, for example, a rotating spotlight in a darkened room. A beam of light will be swept along the walls. The illuminated points on the walls certainly share a common property, but they do not stand in causal relations to each other. Hence they constitute only a pseudo-process.

How is the distinction between causal processes and pseudo-processes to be drawn? Salmon suggests that it can be done by making use of the idea underlying Reichenbach's mark method: a genuine causal process has 'the ability to transmit a mark'; a pseudo-process has not.[24]

But what account can be given of talk about the ability to transmit a mark? Can it be explained without employing concepts of a sort that Hume proscribed? Salmon contends that it can be, and in order to support this claim he sets out what he refers to as an 'at-at' theory of the transmission of a mark, modelled on the familiar 'at-at' analysis of motion in terms of position at different times. This account of what it is for a mark to be *transmitted* may be expressed as follows: 'A mark that has been introduced into a process by means of a single intervention at point A is transmitted

[23] Salmon, 'Theoretical Explanation', *Explanation*, ed. Stefan Korner, Oxford, 1975, pp. 118–43, see p. 125; 'Why Ask "Why?"?', p. 689. Reichenbach, op. cit., pp. 158–61. Suppes, op. cit., pp. 12, 28.

[24] Salmon, 'An "At-At" Theory of Causal Influence', *Philosophy of Science*, 44, 177, pp. 215–24. See p. 220. Also, 'Theoretical Explanation', pp. 129–32.

to point *B* if it occurs at *B* and at all stages of the process between *A* and *B* without additional interventions.'[25]

Is Salmon correct in holding that this account of the transmission of a mark does not violate Humean strictures? In particular, what about the reference to a mark being *introduced* into a process? Does not this involve the notion of causal direction?

This objection is, I believe, correct, but I do not think that it undermines the basic idea. The reason is that it does not really matter whether the event that occurs at point *A* introduces a mark into the process, which is then transmitted to *B*, or whether, instead, it removes a mark that has been transmitted from *B* to *A*, thus giving rise to a mark-free process from point *A* onward. The crucial idea is simply that events occurring at points in a genuine process can result in changes which then persist, and that this does not happen in the case of pseudo-processes. So while talk about introducing a mark is not entirely felicitous, it appears that it should not be difficult to reformulate the analysis to eliminate this feature.

Let us suppose, then, that a satisfactory account can be given of what it is for a mark to be transmitted from one point to another. What about the *ability* to transmit a mark? Can one give an account of that concept which does not violate Humean requirements? Since the concept of an ability presupposes the concept of a law, the answer will depend upon the correct account of the nature of laws. If laws can be identified with regularities, there will, of course, be no problem. A process in which a mark is not transmitted will have the ability to transmit one if it is similar, in relevant respects, to processes in which marks are transmitted.

A feature that causal processes often have, and which Salmon stresses, is that of belonging to a group of causal processes that share a common centre. The causal processes sharing a common centre may be similar, as in the case of those that result when a stone strikes the surface of a pond, or they may vary considerably. The frequent occurrences of causal processes with a common centre has the consequence that certain types of event occur at different places, at the same time, much more frequently than would be the case if their occurrences were independent. This fact led Reichenbach to formulate his principle of the common

[25] Salmon, 'An "At-At" Theory of Causal Influence', p. 221.

cause: 'If an improbable coincidence has occurred, there must exist a common cause'[26]—a principle which Salmon deems very important.[27]

Reichenbach wanted to capture this notion of causal processes dependent upon a central event in a precise way, and in order to do this, he formulated the idea of a *conjunctive fork*. This is a statistical relation among three types of event, *A*, *B*, and *C*—where *A* and *B* are the types of event involved in two causal processes with a common centre, and *C* is the type of the central event to which they are related. Reichenbach suggested that the desired notion of a conjunctive fork could be expressed as follows:

The ordered triple of events (*a*, *c*, *b*) is a *conjunctive fork* if and only if there are event-types, *A*, *B*, and *C* such that *a* is of type *A*, *b* is of type *B*, *c* is of type *C*, and the following conditions are satisfied:

(1) $\text{Prob}(A \,\&\, B, C) = \text{Prob}(A, C) \times \text{Prob}(B, C)$.
(2) $\text{Prob}(A \,\&\, B, \text{not-}C) = \text{Prob}(A, \text{not-}C) \times \text{Prob}(B, \text{not-}C)$.
(3) $\text{Prob}(A, C) > \text{Prob}(A, \text{not-}C)$.
(4) $\text{Prob}(B, C) > \text{Prob}(B, \text{not-}C)$.[28]

Finally, causal interactions. Salmon's suggestion here is that what it is for events to constitute a case of causal interaction can be explained in terms of a different sort of fork—the interactive fork—which can also be defined in terms of statistical relations among events of the types in question. The characterization of an interactive fork which Salmon offers may be expressed as follows:

The ordered triple of events (*a*, *c*, *b*) is an *interactive fork* if and only if there are event-types *A*, *B*, and *C* such that *a* is of type *A*, *b* is of type *B*, *c* is of type *C*, and the following conditions are satisfied:[29]

(1*) $\text{Prob}(A \,\&\, B, C) > \text{Prob}(A, C) \times \text{Prob}(B, C)$

[26] Reichenbach, op. cit., p. 157.
[27] Salmon, 'Why Ask "Why?"?', p. 692. Also, 'Theoretical Explanation', p. 167.
[28] Reichenbach, op. cit., p. 159. Reichenbach's notation for probabilities is slightly different, and he also uses terms which refer sometimes to events, and at other times to types of event.
[29] Salmon, 'Why Ask "Why?"?', pp. 694, 704–5.

(2) Prob(A & B, not-C) = Prob(A, not-C) × Prob(B, not-C)

(3) Prob(A, C) > Prob(A, not-C)

(4) Prob(B, C) > Prob(B, not-C).

The only difference between a conjunctive fork and an interactive fork, then, is with regard to the first condition. It may be helpful to illustrate this difference with examples of each type of fork. Consider two toy boats floating in a pond into which a large rock is dropped. The probability that both boats will be capsized given that a stone of a certain size is dropped in a certain place, from a specified height, etc., will be equal to the probability that the one boat will capsize, under those conditions, times the probability that the other will capsize, under those conditions. The events in question form a conjunctive fork. Contrast this situation with the case, mentioned by Salmon, of the Compton scattering experiment, in which an energetic photon collides with an electron, giving rise to a photon with energy E_1 and an electron with energy E_2. Here the likelihood of this joint result, given the collision, must be much greater than the product of the probability that there will be a photon with energy E_1, given the collision, with the probability that there will be an electron with energy E_2, given the collision. The reason is that, in view of the fact that the principle of the conservation of energy is at least approximately true, there must be a strong correlation between the energy of the photon and that of the electron after the collision: the sum, $E_1 + E_2$, must be roughly equal to the energy of the photon before the collision. Thus the probability that the energy of the electron will be equal to E_2, given the collision *and* given that the energy of the photon is equal to E_1, will be greater than the probability that the energy of the electron will be equal to E_2, given only the collision. And this entails that the likelihood of the joint outcome, given the collision, is greater than the product of the likelihoods of each individual outcome, given the collision. We have, therefore, an interactive fork, rather than a conjunctive one.[30]

Given these accounts of causal processes and causal interactions, how is causal priority to be defined? Salmon's answer, aside from one modification, is essentially along the lines suggested by Reichenbach in section 19 of *The Direction of Time*. The basic idea

[30] Ibid., pp. 692–4.

may be described as follows. One might initially suppose that if (a, c, b) is a conjunctive fork, then c is causally prior to a and to b. But this is not true: c may be an effect of a and b, rather than a cause. However, Reichenbach maintains that it is *true in our world*, first, that if (a, e, b) is a conjunctive fork where e is the common effect of a and b, then there is some event c such that (a, c, b) is a conjunctive fork, and c is the common cause of a and b, and second, that there are cases in which (a, c, b) is a conjunctive fork, and c is the common cause of a and b, and there is no event e such that (a, e, b) is a conjunctive fork and e is the common effect of a and b. This suggests that one should draw a distinction between open and closed conjunctive forks:

A conjunctive fork (a, c, b) is *open*

if and only if

there is no e, distinct from c, such that (a, e, b) is a conjunctive fork.

A conjunctive fork (a, c, b) is *closed*

if and only if

(a, c, b) is not open.

In the case of closed conjunctive forks, there will be—aside from cases of causal overdetermination—two possibilities as to the direction of the causal processes involved. By contrast, if Reichenbach is right with regard to the claims mentioned above, all open forks are, so to speak, oriented in the same causal direction. As a result, one might propose the following definition:

Event c is non-derivatively causally prior to event a

if and only if

there is a conjunctive fork (a, c, b) which is open.[31]

This defines causal priority for events belonging to open conjunctive forks. The next step is to extend it to other pairs of events. This can be done in a number of somewhat different ways.

[31] Compare Reichenbach, op. cit., pp. 161–3. What I have described here as an account of the relation of non-derivative, causal priority, Reichenbach describes as an account of the relation of temporal priority. Reichenbach also speaks of conjunctive forks as being open or closed 'on one side', rather than as simply open or closed. It seems to me that this is only possible after one has a general account of causal priority.

The details, however, are unimportant in the present context.[32]

Salmon's divergence from Reichenbach arises from the fact that Salmon believes that reference to two types of fork is required for a satisfactory account of causal relations. For Salmon, then, the fundamental asymmetry of causation rests upon 'the principle of the common cause—so construed as to make reference to both conjunctive forks and interactive forks'.[33]

Salmon's approach to the problem of providing an account of causation contains much that is very interesting. In the end, however, it does not seem to me to be a satisfactory account. First, there is the question of the connection between causation and positive statistical relevance. As I pointed out, the assumption that, if events of type B are causes of events of type A, then it must be the case that

$$\text{Prob}(A, B) > \text{Prob}(A)$$

is generally accepted by philosophers who favour a probabilistic approach to causation. Nevertheless, I think that it can be shown that this claim should, in fact, be rejected. Suppose, for example that there are the following types of events—$A, B, C, D_1, D_2, \ldots D_n$—which stand in the following relations:

$$\text{Prob}(A, B) = 0.8$$
$$\text{Prob}(A, C) = 0.9$$
$$\text{Prob}(A, D_i) = 0.1, \text{ for } i = 1, 2, \ldots n.$$

Given these probability assignments, whether or not $\text{Prob}(A, B)$ is greater than $\text{Prob}(A)$ will depend upon how $\text{Prob}(C)$ compares with $\text{Prob}(D_i)$ for $i = 1, 2, \ldots n$. If $\text{Prob}(C)$ is not too much greater than $\text{Prob}(D_i)$, for $i = 1, 2, \ldots n$, then $\text{Prob}(A, B)$ will be greater than $\text{Prob}(A)$. On the other hand, if $\text{Prob}(C)$ is sufficiently greater than $\text{Prob}(D_i)$, for $i = 1, 2, \ldots n$, $\text{Prob}(A, B)$ will be less than $\text{Prob}(A)$. The upshot is that if causation entails positive statistical relevance, then if follows that whether or not events of one type cause events of other types will depend, in some cases, upon the relative probabilities of still other types of event. As this

---

[32] Reichenbach sets out one possible approach in op. cit., sects. 4 and 5. For a criticism of his suggestion, together with an alternative approach, see von Bretzel, 'Concerning a Probabilistic Theory of Causation Adequate for the Causal Theory of Time', *Synthese*, 35, 1977, pp. 173–90.

[33] Salmon, op. cit., p. 696.

result seems strongly counter-intuitive, it provides, I think, a good reason for rejecting the claimed connection between causation and positive statistical relevance.

Next, Salmon's account of causal processes: the dubious feature of his analysis would seem to be his insistence that causal processes must exhibit *spatiotemporal continuity*.[34] It is this condition, imposed upon causal processes, which leads him to the conclusion that there is a 'breakdown of causality' in the quantum domain.[35]

It is possible that all the causal processes in our world do have the property of spatiotemporal continuity. But surely it is very implausible to hold that it is part of the *concept* of causal processes that they be spatiotemporally continuous. In the first place, if this were so, one would have to hold that those dualist theories of the mind according to which mental events have no spatial locations, but do enter into causal relations, either with physical events or other mental events, were not merely incoherent, but transparently so. Secondly, consider the possibility of gappy laws in physics. Suppose that the world were completely deterministic, but that the transmission of light waves involved (completely predictable) jumps or gaps. Would we not still regard the propagation of light waves as a paradigm case of a causal process? Or suppose that it turned out that the law of gravitation involved discontinuous functions, so that while the force field acting upon any sufficiently large region of space had a magnitude very close to that specified by an inverse square law, in very small sub-regions there would be either a much larger force, or none at all. If this were the case, would one cease to hold that the acceleration of the earth as it moved in a roughly elliptical orbit around the sun was caused by the presence of the sun? I do not think that there is any reason at all to believe that one would. Consequently, it seems that any analysis of causation which entails that causal processes must exhibit spatiotemporal continuity cannot be adequate.

Why does Salmon impose this requirement? His reason appears to be connected with one of the points on which he diverges from Reichenbach. As was mentioned above, Reichenbach thought that causal relations could be explicated in terms of statistical relations among the types of event in question. Salmon does not. He holds

---

[34] Salmon, 'Theoretical Explanation', pp. 128 ff., and 'Why Ask "Why?"?', p. 690.
[35] Salmon, 'Theoretical Explanation', p. 133.

that one 'must also establish appropriate connections via causal processes'.[36] His rationale, then, for insisting upon the continuity requirement appears to be that if it were dropped, causal processes would have to be explained in purely statistical terms, and one would therefore be back with the unsatisfactory view espoused by Reichenbach. But we shall see, in the next chapter, that this is not in fact the case.

There are also questions that can be raised concerning Salmon's attempt to distinguish between causal processes and pseudo-processes by reference to the ability to transmit a mark. If, for example, a singularist conception of causation is correct, then Salmon's account will not succeed in correctly classifying causal processes that involve causal connections for which there are no underlying laws. But I think it is best to waive this point, for even if one accepts a singularist conception, one can view Salmon's account as attempting to deal only with causal processes that are governed by underlying laws.

Salmon points out, in a later article, that the account of mark transmission offered in 'An "At-At" Theory of Causal Influence' is not entirely satisfactory, but he expresses the view that the defect can be remedied.[37] I am inclined to think that Salmon is right. But even if the defect cannot be repaired, this need not be a serious matter for Salmon's general approach. For it would seem that one should be able to offer, for law-governed causal processes, an alternative account of the distinction between causal processes and pseudo-processes, based upon the different relations between events and underlying laws in the two cases.

Next, Salmon's account of causal interactions: here I shall confine myself to a very brief observation, but one that points to an important condition which it would seem that any analysis of causation should satisfy. Salmon's explanation of the concept of causal interactions involves the notion of an interactive fork. One condition that an interactive fork must satisfy, and the one which distinguishes interactive forks from conjunctive forks, is that the corresponding event-types—A, B, and C—must be such that:

$$(\text{1*}) \ \text{Prob}(A \ \& \ B, C) > \text{Prob}(A, C) \times \text{Prob}(B, C).$$

Consider, now, a completely deterministic world, in which there

[36] Salmon, 'Why Ask "Why?"?', p. 705.
[37] Ibid., pp. 690, 703.

are interactions—involving events of types A, B, and C—where an event of type C is causally *completely sufficient* for the occurrence of events of types A and B. In that case, Prob(A & B, C), Prob(A, C), and Prob(B, C) will all be equal to one, so condition (1^*) cannot be satisfied.

The emphasis which Salmon, together with Reichenbach and Suppes, has placed upon the desirability of an account of causation which admits the possibility of what might be called probabilistic causal relations, seems to me very important. But as the above point illustrates, care must be taken to avoid the opposite error— that of offering an account which admits probabilistic causal relations, but leaves no room for completely deterministic ones. Salmon's account of causal interaction appears to suffer from this defect.

This brings us to the crucial issue: the analysis of causal priority. Salmon suggests that this notion can be explicated in terms of conjunctive and interactive forks. This idea is exposed, however, to a number of serious objections—including the three that were advanced, in the preceding section, against attempts to analyse causal priority in terms of entropic or non-entropic irreversibility.

The first objection is this. Suppose that event a is causally connected with event b. What determines the direction of that connection? The crucial point is that, on Salmon's approach, it is *not* ontologically settled by the relevant laws, together with that part of the world that contains events a and b, together with the events that are causally intermediate. This is particularly clear if events a and b belong to some causal process, or to a closed fork— since the direction of causation is determined by open forks. This dependence of the direction of the causal connection between two events upon facts about other events that are not causally intermediate is surely very counter-intuitive.

Secondly, given Salmon's approach, no direction for causation will exist at all unless *all* open forks face in the same direction. This implies that whether one event caused another will only be ontologically settled by the total history of the world.

Thirdly, there is the 'inverted worlds' objection. Suppose that it is true in our world that all forks that are open are open towards the future, and hence in the direction of causation. Then, in the inverted, twin world, all open forks will be open towards the past, and hence in the direction opposite to that of causation. The

possibility of inverted worlds therefore shows that the direction of
causation cannot be defined in terms of the direction determined
by open forks.

Fourthly, Salmon's approach is open to a 'simple world'
objection. Consider a world in which (1) the only forces that are
nomologically possible are contact forces, and (2) where it
happens that the trajectories of the particles are such that there are
never any collisions. Given that the motion of a particle is a causal
process, such a world will contain causal processes, and they will
have a direction, even though the world is totally devoid of both
conjunctive and interactive forks. Salmon's account of the
direction of causation is not compatible with this possibility.

The above objections show that the analysis proposed by
Salmon does not succeed in capturing the *concepts* of causal
priority and causal direction. The thrust of my final objection is
that, even if one were to set aside the problem of providing an
analysis of causal concepts, and attempt only to formulate
concepts that would be *extensionally equivalent*, the account
offered would still have to be rejected.

This final objection turns upon the point that Salmon's account
presupposes that *all* open forks are open in the same direction,
namely, towards the future. If this is not the case, the account is
unsatisfactory. The question, therefore, is whether there is reason
to believe that all open forks are open to the future, and none to
the past. If one considers rather complex forks, involving a large
number of particles–such as that involved when a stone is dropped
into a pond producing outgoing concentric waves—it may seem
rather plausible that all open forks are probably open to the
future. But not all forks are very complex—as is illustrated by the
case of the Compton scattering effect. And once such simple forks
are considered, the situation appears very different. For when
simple forks are taken into account, it is surely most unlikely that
there are *no* forks, either conjunctive or interactive, that are open
to the past. There is, then, excellent reason for thinking that the
account advanced by Salmon does not provide concepts that are
satisfactory even from a purely extensional point of view.

7.6 CONTROLLABILITY AND CAUSAL PRIORITY

The final approach to causal priority which I wish to consider attempts to relate it to the notion of possible control. An approach of this sort has been advanced by von Wright, among others:

> I now propose the following way of distinguishing between cause and effect by means of the notion of action: *p* is a cause relative to *q*, and *q* an effect relative to *p*, if and only if by doing *p* we could bring about *q* or by suppressing *p* we could remove *q* or prevent it from happening.[38]

One common objection to this type of account is that, as it stands, it limits causal relations to states of affairs that can be manipulated. The account can, of course, be extended to cover causal relations between events that cannot in fact be so controlled—for example, by introducing reference to *possible* agents. But even when this is done, the paradigm case of causation will still be that involving personal action, and the concept of action will still be semantically prior to the concept of causal priority. And this seems quite counter-intuitive. To talk about causal relations between states of affairs does not seem to be to talk about what agents, either actual or hypothetical, could do.

There is, however, an even more fundamental objection, namely, that the analysis, as it stands, is circular, since at least one of the concepts employed in the analysans presupposes the concept being analysed. In particular, the analysans involves the concept of bringing about *p* by doing *q*, and this surely entails that *p* is causally prior to *q*.

This flaw has sometimes gone unnoticed. The explanation, perhaps, is that there is a tendency, in considering examples of bringing about one thing by means of another, to make tacit appeal to temporal considerations. As long as this appeal remains at the tacit level, the concept of bringing about one thing by means of another may seem unproblematic.

As an illustration consider the case of a match which, if dry, in the presence of oxygen, etc., lights if struck with sufficient force. One can light the match by striking it when it is dry etc., but one cannot bring it about that it was not struck by bringing it about that it did not light, even though it was dry etc. If asked why the second

---

[38] Von Wright, *Explanation and Understanding*, Ithaca, 1971, p. 70.

claim is to be rejected, it is natural to respond that to bring it about that it did not light, even though dry etc., one must *first* bring it about that it wasn't struck. But a difficulty emerges when one asks how 'first' is to be interpreted here. If it means temporal priority, then the account of causal priority in terms of controllability will entail that one state of affairs can be causally prior to another only if it is also temporally prior. The controllability account will then be exposed to the same objections as the account that equates causal priority with temporal priority. And on the other hand, if 'first' means causal priority rather than temporal priority, the account of causal priority in terms of controllability is circular.

Not all philosophers who have advocated analysing causal priority in terms of the concept of control have been oblivious, however, to the need to offer an analysis of what it is to bring about one state by means of another. Gasking, for example, is very aware of the need for such an analysis, and he summarizes his own approach to the problem as follows:

I have made two points:

First: that one says '*A* causes *B*' in cases where one could produce an event or state of the *A* sort as a means to producing one of the *B* sort. I have, that is, explained the 'cause–effect' relation in terms of the 'producing-by-means-of' relation.

Second: I have tried to give a general account of the producing-by-means-of relation itself: what it is to produce *B* by producing *A*. We learn by experience that whenever in certain conditions we manipulate objects in a certain way a certain change, *A*, occurs. Performing this manipulation is then called: 'producing *A*'. We learn also that in certain special cases, or when certain additional conditions are also present, the manipulation in question also results in another sort of change, *B*. In these cases the manipulation is also called 'producing *B*', and, since it is in general the manipulation of producing *A*, in this case it is called 'producing *B* by producing *A*'. For example, one makes iron glow by heating it. And I discussed two sorts of cases where one does not speak of 'producing *B* by producing *A*'. (1) Where the manipulation for producing *A* is the general technique for producing *B*, so that one cannot speak of 'producing *B* by producing *A*', but only *vice-versa*. (2) Where the given manipulation invariably produces both *A* and *B*, so that the manipulation for producing *B* is not a special case only of that for producing *A*.[39]

Gasking's proposal is an interesting one. His analysis is,

[39] Gasking, op. cit. See pp. 485–6.

however, open to at least three objections. In the first place, the analysis is not sufficiently general, since it provides no account in the case where a state of affairs of type A is both nomologically necessary and nomologically sufficient for a state of affairs of type B. The above passage suggests that Gasking probably holds that, in such a case, one cannot produce states of type B by means of states of type A, or vice versa. But this view is surely very counter-intuitive. Why should it not be possible for it to be the case both that one can control states of type A, and that states of type A are both causally necessary and causally sufficient for states of type B? And if this is possible, then it must be possible to bring about one type of state by means of another even when it is nomologically impossible for either type of state to exist without the other.

In the second place, not all cases that satisfy the suggested account are cases where a state of type B is brought about by a state of type A. For suppose that doing M always produces a state of type A, but produces a state of type B only under conditions C. There are then two very different possibilities. One is that states of type A, combined with conditions of type C, causally give rise to states of type B. The other is that doing M under conditions of type C *directly* gives rise to a state of type B. Gasking does not distinguish between these two cases. Yet it is only in the former case that one can speak of producing B by means of A.

In the third place, the analysis of causation offered still suffers from circularity, since the concept of manipulation appears to involve causal concepts, possibly in two ways. First, manipulation is a type of action, and it seems quite plausible that action necessarily involves a causal relation between mental states and the states brought about by the action. It is possible, of course, to argue that what is involved here is not ordinary causation but something different. I think it is very doubtful that this contention can be sustained. But even if it can be, there will still be a problem since the concept of manipulation also involves the notion of a causal connection between bodily movements and changes in objects in contact with one's body. So it would seem that any account of causal priority that involves the concept of manipulation will suffer from circularity.

What are the prospects for offering a revised analysis which will surmount these difficulties? With regard to the first two problems, the chances may be fairly good. But I think that the situation is very different with regard to the third. The concepts of controlling, manipulating, producing, bringing about by means of, and so on, all involve the concept of causal priority, and appear to be semantically more complex than that concept. As a consequence it seems inescapable that any attempt to offer an analysis of causal priority in terms of such concepts will suffer from circularity.

There is, however, a related possibility that deserves to be considered. What if one abandons the notion of control in favour of the weaker notion of *apparent control*? The point is that to say that one thing is apparently controlling another does not imply that they are causally related. This means that if one could give an analysis of the direction of causation in terms of the direction of apparent control, the result would be an account that was in the spirit of the approach we have been considering, but that avoided the problem of circularity.

This alternative would, however, encounter problems of its own—depending upon the precise account offered of the concept of apparent control. An analysis according to which apparent control entailed temporal priority would, for example, face the difficulties discussed in section 7.1. Let us suppose, then, that one state of affairs can stand in the relation of apparent control to another without being temporally prior. The resulting analysis of the direction of causation in terms of the direction of apparent control would then fall prey to two other objections mentioned above. First, there is the 'simple world' objection. Consider again the case of the universe that consists of two neutrons rotating endlessly about one other. If the concept of apparent control involves neither causal concepts, nor the concept of temporal priority, then it seems clear that there is no direction of *apparent* control in that world—though there certainly is a direction of actual control. Therefore the direction of causation cannot be identical with the direction of apparent control.

Second, there is the 'inverted worlds' objection. The direction of apparent control in our world runs, for example, from events which occur in the year 1990 to events which occur in the year 2000. If the notion of apparent control involves neither causal concepts, nor the concept of temporal priority, then, in the inverted twin-world, the direction of apparent control will run from 1990*-events to 2000*-events, where 1990*-events are those that exactly correspond, except for a reversal of all velocities, to those events in our world that happen in 1990, and similarly for 2000*-events. But the direction of causation in the inverted world is, by hypothesis, from 2000*-events to 1990*-events. So the direction of causation in the inverted world is opposite to that of the direction of apparent control.

7.7 CONCLUSIONS

I have tried to survey the most important attempts to provide a solution to the difficult, and crucial problem, of causal priority. If the arguments set out above are sound, none of the accounts is completely satisfactory. This does not, of course, mean that the considerations appealed to in these accounts are of no relevance to the construction of an adequate account of causal priority. Consider, for example, the direction of irreversible processes, both entropic and non-entropic, or the direction of open forks, or the direction of apparent control. Though none of these are constitutive of causal priority, it seems to me that all of them provide excellent evidence for beliefs concerning the direction of causal processes. Any adequate account of causal priority will need to be able to provide an explanation of these evidential relations.

Secondly, the arguments which were employed above against the various accounts of causal priority do more than show that certain accounts are unsatisfactory. They also support some general conclusions—some made explicit in the above discussion, others not—which impose rather strong constraints upon the analysis of causal concepts. In the next two chapters, after setting out a realist account of causation, I shall touch upon some of those constraints; for I believe that one of the merits of a realist approach to causation is its compatibility with the constraints in question.

─── 8 ───

The Nature of Causation:
The Basic Approach

My object, in this chapter and the next, is to put forward a detailed account of the nature of causation. I shall begin by advancing two very general theses concerning any adequate account of causation. The first is that only a *realist* account can possibly be satisfactory. The second is that any acceptable account will have to treat causal relations as *theoretical* relative to *all* non-causal facts. Having developed these general theses, I shall then attempt to set out, and to defend, a precise account of the nature of causation.

8.1 CAUSATION: A REALIST ACCOUNT

The choice between realist and non-realist views on the existence of theoretical entities has been much discussed in philosophy of science, and many philosophers accept a realist view. The situation, in contrast, is very different in the case of causation. Here the question of the choice between realist and non-realist accounts is rarely given much attention, and it seems that very few analytically-oriented philosophers are willing to embrace a realist account.

  The failure to consider this issue seriously seems to me unfortunate, for I believe that one reason that the problem of the nature of causation has proved so intractable is that realist accounts have not been canvassed. In this present section I shall attempt to indicate why it seems to me likely that only a realist account of causation can possibly be correct.

8.1.1 *Causal Realism*

What constitutes a realist account of causation? This question is perhaps best answered by considering the much more familiar contrast between realist and non-realist views with respect to the existence of theoretical entities. What is it to be a realist in this sphere? It would seem to involve the acceptance of two theses—a semantical thesis, and an epistemological one. To begin with the semantical thesis, there are three general views that one can take concerning the meaning of statements containing theoretical terms. One is that theoretical terms cannot possibly refer to anything, and that statements containing such terms cannot be either true or false. This is the view accepted by the instrumentalist, who maintains that theories do not describe reality, that they merely function as devices which, given observation statements as input, produce other observation statements as output. A good theory will be one that generally produces true statements when true statements are put in, and which has various other virtues such as comprehensiveness and simplicity.

The second possible view is that theoretical terms can refer, and statements containing such terms are either true or false, but that theoretical terms do not refer to anything over and above what is observable, so that what makes a theoretical statement true is nothing distinct from observable facts. According to this reductionist view it may be true, in a sense, that theoretical entities exist, but they will not possess the status of logically independent entities.

The third view also involves the claims that theoretical terms can refer, and that statements containing such terms are either true or false. But in addition, it is maintained that theoretical terms can refer beyond what is observable, with the consequence that the truth-values of theoretical statements are not determined by observable facts. According to this final view, theoretical entities can be said to exist without any qualification, since their existence is not logically supervenient upon any observable states of affairs.

To be a realist with regard to theoretical entities involves accepting this third view. That, however, is not sufficient. For as Bas van Fraassen has emphasized, one might agree with the realist concerning the correct interpretation of statements containing theoretical terms, yet hold that it is not possible, even in principle, to have good reasons for believing that any particular theoretical statement is true. This leads to the following view of science. On the one hand, it may be impossible to construct satisfactory theories without employing terms that purport to refer to what is unobservable. But on the other, it cannot be a necessary condition for being a good scientific theory that there be good reasons for believing that it is true. A more modest requirement is needed, namely, that there be good reasons for believing that the theory is empirically adequate—where this is essentially a matter of agreement with respect to what is observed.[1]

The upshot is that in order to be a realist, one must accept both a semantical thesis and an epistemological thesis. There is some question as to how exactly the epistemological thesis should be formulated. However, a rather plausible view is that it should involve at least the claim that it is possible in principle to have observational evidence that would make it likely that some particular theoretical statement is true, and which would therefore provide one with good reason for believing that certain sorts of theoretical entity do in fact exist.

What it is to be a realist with respect to causation can now be explained. It too involves accepting both a semantical thesis and an epistemological one. The semantical thesis is straightforward. It involves, first, the claim that causal statements are either true or false, and secondly, the claim that the truth-values of causal statements are not, in general, logically determined by non-causal facts.

What about the epistemological thesis? If one paralleled the formulation used in the case of realism with regard to theoretical entities—merely replacing the terms 'theoretical' and 'obser-vational', respectively, by the terms 'causal' and 'non-causal'—one would have the claim that it is possible in principle to have non-causal evidence that makes it likely that some causal statement is true. This formulation, however, is unsatisfactory. For some philosophers who are causal realists would hold that it is possible to have *non-inferential* knowledge of causal connections, and thus that it is not necessary to claim that knowledge concerning non-causal facts can constitute evidence for causal beliefs.

This brings out a rather important point, namely that there are two very different types of causal realism. According to the one view, causal discourse is semantically autonomous and cannot be analysed in terms of non-causal expressions. As a consequence, if

---

[1] Van Fraassen, *The Scientific Image*, Oxford, 1980, chaps. 1–3.

one is to have any knowledge of causal states of affairs, at least some of it must be non-inferential. According to the other view—which I shall espouse—causal relations are theoretical relations and, as a consequence, any knowledge that we have of causal connections must ultimately rest upon knowledge of non-causal states of affairs.

What is needed, therefore, is a formulation of the epistemological thesis which does not discriminate against either of these realist positions. The following seems plausible: it is possible in principle to be rationally justified in accepting some causal statement as true.

To sum up then, a realist view of causation involves, first, the semantical claims that causal statements are either true or false, and that their truth-values are not logically determined by non-causal states of affairs, and second, the epistemological claim that it is possible in principle to be rationally justified in believing that particular causal statements are true.

8.1.2 *The Case for Causal Realism*

Most philosophers accept a reductionist view of causation, rather than a realist one. Why is this so? One possible explanation is that philosophers tend to equate causal realism with those versions of it which involve the claim that causal expressions cannot be analysed in terms of non-causal ones. As this claim strikes most philosophers as unappealing, causal realism is rejected.

What reasons can be offered for rejecting the claim that causal expressions cannot be analysed in non-causal terms? One argument was offered in section 1.3. The thrust of that argument was, first, that analytically basic descriptive terms must apply to things by virtue of observable properties and/or relations; second, that there are possible worlds that are observationally indistinguishable, but where one involves causally related events, while the other does not; and therefore, thirdly, that causal expressions must be analysable in non-causal terms.

This is the argument on which I would rest the most weight. There are, however, at least three other arguments that deserve to be mentioned. The first of these parallels an argument set out in section 7.1 in support of the thesis that the concept of temporal priority cannot be analytically basic, and it runs as follows: All causal relations appear to possess formal properties. The relation of causation which obtains between states of affairs, for example, is necessarily irreflexive, necessarily asymmetric, and necessarily transitive. How are these necessary structural properties to be explained? The only clear possibility, it would seem, involves showing that the relevant statements can be derived from logical truths in the narrow sense by the substitution of synonymous expressions. Accordingly, if one is to have an intelligible account of why it is that causal relations necessarily possess the formal properties that they do, it would seem that causal concepts must be analysable.

The second argument involves an attempt to show that if causal concepts are unanalysable, then it follows that the acquisition of causal knowledge presupposes the existence of a somewhat mysterious capacity. The argument may be developed as follows. Assume that causal concepts cannot be analysed in non-causal terms. How, in that case, is one ever able to learn the meaning of causal terms? The only answer, it would seem, is that one must possess a capacity for being directly aware of causal relations between states of affairs. This direct awareness provides one with non-inferential knowledge of causal connections, and so makes possible the acquisition of causal concepts.

It is not easy to judge how much force there is in this line of thought. For there may be ways of supporting the counter that such a faculty is not really mysterious. One way, for example, is to appeal to the idea that we are directly aware of temporal relations between events, and then to argue that the direction of time, at least, must be analysed in terms of the direction of causation. If these two claims could be sustained, it would have been shown that we can be directly aware of the direction of causal processes. Direct awareness of causal processes would be no more mysterious than direct awareness of temporal relations between events.

This particular line of argument is not, of course, available to anyone who wishes to hold that the direction of causation is to be analysed in terms of the direction of time. However, there may be other ways in which one could try to show that there is nothing objectionable in the claim that at least some of our knowledge of causal relations is non-inferential.

The third and final argument turns upon the claim that non-causal facts can provide evidence for causal claims, and it can be set out as follows. Suppose that a continuous film is made of some occurrence, that the film is later divided up into frames, and that these are then thoroughly shuffled. It will, in general, be possible for someone who did not witness the occurrence to work out the correct order of the frames, and if the occurrence involved irreversible processes, it will also be possible to determine the direction of causation. This would seem to show that information which is not itself causal can provide very good evidence for causal hypotheses. The question now is whether any satisfactory account of such evidential relations can be given if causal concepts cannot be analysed in non-causal terms.

A possible response to this argument is that when one draws inferences such as the above, one is tacitly making use of previously acquired background information about the direction of causation in processes of the relevant sort. Can this defence be sustained? It seems to me to be very problematic. For consider a person who has never observed the propagation of concentric waves. If such an individual were asked to compare the possibility of spontaneously generated outgoing waves with that of sponta-neously generated incoming waves, does it not seem plausible that he could arrive at the conclusion that the former are immensely more probable than the latter, simply by thinking about the two situations? If this can be done, then it would seem that background information cannot be necessary for the inference described above. There may be a serious problem, therefore, in showing how certain sorts of non-causal evidence can bear upon causal hypotheses, if one accepts the thesis of unanalysability.

In view of the above arguments, there would seem to be good grounds for rejecting the thesis that causal concepts cannot be analysed in non-causal terms, and, along with it, any version of causal realism that incorporates that unanalysability claim. But it does not follow from this that causal realism is untenable, since, as was pointed out above, causal realism may take another, very different form. According to this other type of causal realism, causal terms are not unanalysable. For they are theoretical terms, and can therefore be analysed by means of any method that is applicable to theoretical terms in general—such as the Ramsey/Lewis approach discussed in Chapter 1. Moreover, given that causal terms are theoretical, there is no need to postulate any special faculty that provides one with direct awareness of causal relations. Causal claims will be epistemologically justified in the same ways that theoretical claims in general are justified.

The upshot is that the type of causal realism that I shall be advancing does not possess the unappealing features characteristic of other versions of causal realism. In the absence of further arguments, therefore, there does not appear to be any justification for thinking that a realist approach to causation is unsound. But are there positive reasons for thinking that it is worth pursuing? I believe that there are. In the first place, realist approaches to causation have hardly been considered, let alone seriously explored. In contrast, a great deal of philosophical effort has been devoted to the attempt to find a satisfactory reductionist account. But in spite of many ingenious suggestions, nothing even approximating an acceptable account has been advanced. In view of this, there would seem to be good reason for exploring realist alternatives.

Secondly, consider the possibly closely related question of the nature of laws. Here, too, philosophical energy has been expended mainly on attempts to develop a satisfactory reductionist account. But we saw in Part II, first, that reductionist accounts of laws are exposed to very serious objections, and second, that there is a promising realist alternative. If a realist account of laws is required, isn't that a reason for thinking that a realist approach to causation may also be needed?

The third and the most important point concerns the relation between a reductionist view of causation and the supervenience view of causal relations. The latter involves the claim that the truth-values of statements concerning causal relations between particular states of affairs are logically determined by what laws there are, together with what non-causal facts there are involving particulars. This thesis does not entail a reductionist view of causation, since the laws which, on the supervenience view, enter into the determination of causal connections between particulars, may involve causal laws. But on the other hand, if a reductionist view of causation is correct, then so is the supervenience view of causal relations. As a consequence, any reason for rejecting the supervenience view of causal relations must also be a reason for rejecting any reductionist account of causation. The arguments set out in Chapter 6 are, therefore, also arguments against any reductionist approach to causation.

8.2 THE UNDERLYING INTUITION

The account of causation that I shall advance involves three fundamental theses. The first is that causal relations are *theoretical* relations between states of affairs. The second is that only a *realist* interpretation of the theoretical terms in question will do. Given these claims, the question is how the relevant theoretical relation is to be picked out or specified. The answer is provided by the third thesis, which can be expressed informally by means of the following slogan: *causation is that theoretical relation that determines the direction of the logical transmission of probabilities.*

The main task to be undertaken in the remainder of this chapter will involve showing how this slogan can be translated into a precise account of the nature of causation. First, however, I think it is very important to grasp the intuition that underlies the account, and which suggests the above characterization. This can be done by considering the following very simple case. Imagine that one has an unbiased coin, and an unbiased die, so that when the coin is flipped, in a random fashion, the probability that it will land heads is one-half, while the probability that the die will come up six when it is rolled, in a random fashion, is equal to one-sixth. Suppose, however, that a special type of room is discovered, such that if the coin is flipped, and the die rolled, in that sort of room, and they come to rest apparently simultaneously, then the coin comes up heads *when and only when* the die comes up six. If we assume that the events are causally connected, what is the direction? Is it the coin's coming up heads, together with the properties of the room, that cause the die to come up six? Or is it the die's coming up six, together with the relevant properties of the room, that cause the coin to come up heads?

As the situation has been described up to this point, there is no reason for choosing one answer rather than the other. But now suppose that you are given some addition information. First, suppose that you learn that in this situation the die comes up six approximately *half the time*. Then it would seem to be very reasonable to conclude that it is the coin's coming up heads, together with the features of the special sort of room, that causally brings it about that the die comes up six. Alternatively, suppose that the additional information is, instead, that in this situation the coin comes up heads only about *one-sixth* of the time. Then the reasonable conclusion would surely be that it is the die's coming up six, together with the special features of the room, that brings it about that the coin comes up heads.

In the first case, the probability of the coin's coming up heads gets 'transmitted', so to speak, to the die's coming up six, and I am suggesting that it is this that leads one to conclude that it is the coin's coming up heads that, together with other factors, causes the die to come up six. In the second case, on the other hand, the transmission of probabilities is in the opposite direction. The probability of the die's coming up six is transmitted to the coin's coming up heads. One therefore concludes that it is the die's coming up six that causes the coin to come up heads.

8.3 A FORMAL EXPLICATION

The intuitive idea is that causation is a theoretical relation that determines the direction of the logical transmission of probabilities. How can this idea be made precise? The task in this section will be to provide an answer to that question. This will involve formulating postulates that specify the ways in which causal relations enter into the determination of the likelihoods of states of affairs, and that can be used to set out a formally adequate theory of causation.

8.3.1 *Two Alternatives*

There are a number of causal concepts. Some are concepts of relations between states of affairs (or events). One state of affairs causes another. Or it is causally sufficient in the circumstances for another. Or it is causally necessary in the circumstances for the other. Or one state of affairs is, by itself, causally completely sufficient for another. Or it is causally completely necessary for the other.

Other causal concepts are of relations between, not states of affairs, but *types* of state of affairs. Thus, for example, one type of state of affairs may be causally sufficient to ensure the existence of a state of affairs of some other type. Or it may be causally necessary for the existence of a state of affairs of some other type.

Some of these concepts, it would seem, can without much

difficulty be defined in terms of other. What it is for one state of affairs to be causally sufficient in the circumstances for another can be explained in terms of what it is for one state of affairs to be, by itself, causally completely sufficient for another, and similarly for the relation of being causally necessary in the circumstances for another state of affairs. Causal sufficiency, as a relation between states of affairs, can be explained, it would seem, in terms of the corresponding relation between types of state of affairs. Finally, what it is for one type of state of affairs to be causally necessary for another can be explained in terms of the relation of causal sufficiency: the possession of property P is causally necessary for the possession of property Q, for example, if and only if the non-possession of property P is causally sufficient to ensure the absence of property Q.

What causal relation (or relations) should be taken as fundamental? The answer will depend upon what view should be taken on the issue discussed in Chapter 6. If the supervenience view is correct, so that causal relations between states of affairs reduce logically to non-causal facts plus causal laws, then the fundamental concept will be that of a causal law, and the basic problem will be to provide a satisfactory account of the nature of causal laws. Once that problem is solved, it will be a relatively straightforward matter to explain most other causal concepts. For one can begin by explaining the relation that obtains when states of affairs of one type are causally sufficient to ensure the existence of states of affairs of some other type in terms of the concept of a causal law, and then go on to explain virtually all other causal concepts along the lines indicated immediately above. However, some difficulty does arise when one comes to the relation that obtains between two states of affairs when one causes the other. How this should be analysed in terms of causally necessary and/or causally sufficient conditions is a matter of dispute. The best-known attempt at providing an answer is that of John Mackie who suggested, in his article 'Causes and Conditions', that at least part of what is generally being claimed in singular causal judgements is that a certain state of affairs is an 'inus' condition, where this is defined as 'an *insufficient* but *necessary* part of a condition which is itself *unnecessary* but *sufficient* for the result'.[2] The details, and possible

---

[2] Mackie, 'Causes and Conditions, *American Philosophical Quarterly*, 2/4, 1965, pp. 245–64. See p. 245.

difficulties, in formulating this sort of analysis of the relation of causation need not concern us here however.

The overall programme of analysis will, in contrast, be very different if the supervenience view has to be abandoned. If, as the arguments advanced in Chapter 6 appear to show, two situations may differ causally even though they differ neither with respect to non-causal properties and relations nor with respect to causal laws, then the relation of causally bringing about must be taken as involving a genuine relation beween particular states of affairs—though it is possible that this relation also depends upon the existence of underlying laws. Thus, rather than attempting to analyse the relation of causally bringing about in terms of causally sufficient and/or causally necessary conditions, or in term of causal laws, one may want to run the analysis in the opposite direction, using the concept of the relation of causation to explain what it is to be a causal law. This is certainly the approach that one will want to follow if one opts for a singularist conception, and it *may* also be the correct approach if one rejects both the supervenience and singularist views in favour of the third way. In any case, one natural suggestion, if one were to adopt this second approach, would be that for it to be a causal law, for example, that if something has property P then it has property Q, is for it to be a law that something's having property P causally brings it about that it has property Q.

In short, we are confronted with two very different alternatives. One involves setting out an analysis of the concept of a causal law, and then analysing other causal concepts in terms of that. The other involves explaining the relation of causally bringing about, and then analysing other causal concepts, including the concept of a causal law, in terms of that.

Which programme should be followed depends upon which of the three possible views on the relation between causal laws and causal relations is correct. Now, in Chapter 6, I tried to show that there are strong arguments against the supervenience view. None the less, it would seem unwise to premise the present discussion upon that conclusion. What I shall do, therefore, is to attempt to formulate three different versions of the general approach to causation that I favour, corresponding to the different views that can be taken on the relation between causal laws, and causal relations between states of affairs.

8.3.2 *Version One: The Supervenience View*

In this section I shall attempt to show how the intuitive idea that causal facts are theoretical ones that determine the direction of the logical transmission of probabilities can be made precise when one adopts the supervenience view of causal relations. Since the fundamental concept is, then, that of a causal law, the task will be to formulate postulates that specify the ways in which causal laws enter into the determination of the likelihoods of states of affairs. The theory constructed from these postulates can then be used to define, via a Ramsey/Lewis approach, what it is to be a causal law.

The discussion in the present section will be restricted to the case of deterministic causal laws. Probabilistic causal laws will be considered in section 9.1.

It will be convenient to have a brief way of expressing causal laws. I shall use the following symbolism:

$$P \to Q$$

whose intended interpretation is:

> It is a causal law that if something has property P, then it has property Q.

At first glance this symbolism may seem unsatisfactory. For while it does enable one to express causal laws to the effect that if something has one property, then that very same thing has some other property, what about causal laws connecting the properties of different objects? In fact, however, the symbolism suggested does not impose any serious limitations with regard to what causal laws can be expressed. For what a causal law will typically involve is not merely that if something has property P, then there will be something else which has some property Q. It will also specify the relation, R, between the cause and the effect. The relevant causal laws will therefore be expressed by statements of the form: 'It is a causal law that, for any x, if x has property P, then there exists a y such that y has property S, and x stands in relation R to y.' But this, in turn, is logically equivalent to the statement: 'It is a causal law that, for any x, if x has property P, then x has the relational property of being such that there is a y that has property S, and such that x stands in relation R to y.' As a result, since the term Q may stand for relational properties as well as for non-relational

256 *Causation*

ones, there is no problem about expressing causal laws connecting properties of different objects by means of the symbolism $P \to Q$.

We shall also need to refer to probability relations. I shall use the expression

$$\text{Prob}(Px, E) = k$$

whose intended interpretation is:

> The logical probability that x has property P, given only evidence E, is equal to k.

Let us now turn to the central task. What I want to suggest is that what is absolutely fundamental to the concept of causation is the idea that the probability of a given state of affairs is a function of the probabilities of states of affairs upon which it is causally dependent, in a way in which it is *not* a function of the probabilities of states of affairs that are causally dependent upon it, either directly or indirectly. What this means, given the supervenience view of causal relations, is that a given causal law has a bearing upon the likelihood of states of affairs of the sort that fall under the consequent of the law, that it does not have with respect to states of affairs that fall under the antecedent.

How is this intuitive idea to be captured in a precise way? One thing that is needed, presumably, is postulates that specify the values of the following three probabilities:

1. $\text{Prob}(Px, P \to Q)$
2. $\text{Prob}(Px, P \to P_1 \& P_1 \to P_2 \& \ldots \& P_n \to Q)$
3. $\text{Prob}(Qx, P \to Q)$.

For the first two represent the likelihood of a state of affairs relative to information about causal laws specifying its effects, either direct or indirect, while the third represents the likelihood of a state of affairs relative to information concerning causal laws that specify types of state of affairs upon which it may be causally dependent.

The postulates which, I believe, specify these values correctly are the following:

(1) $\text{Prob}(Px, P \to Q) = \text{Prob}(Px, L)$
(2) $\text{Prob}(Px, P \to P_1 \& P_1 \to P_2 \& \ldots \& P_n \to Q) = \text{Prob}(Px, L)$

(3) $\text{Prob}(Qx, P \rightarrow Q) =$
$\text{Prob}(Px, L) + \text{Prob}(\sim Px, L) \times \text{Prob}(Qx, \sim Px \,\&\, P \rightarrow Q),$
where L is any logical truth.

Postulate (1) asserts that the likelihood that something will have some property, P, given the information that it is a law that something's having property P causes it to have some other property Q, does not differ from the a priori probability of its having property P.

Postulate (2) asserts, in effect, that the situation is unchanged given the information that there are two or more laws that entail that something's having property P indirectly causes it to have property Q.

Postulate (3), in contrast, asserts that the probability of something's having some property, Q, given that it is a law that something's having property P causes it to have property Q, will be a function of certain other probabilities, including the probability of that thing's having property P. One might expect, therefore, that the probability of something's having property Q, given that it is a law that the possession of P brings about the possession of Q, will generally differ from the a priori probability of its having property Q.

These three postulates deal with the relevant probabilities in cases where one's information is very restricted. One would prefer to have postulates that possess greater generality. Consider, then, the following possibilities:

(1*) $\text{Prob}(Px, P \rightarrow Q \,\&\, S) = \text{Prob}(Px, S)$

(2*) $\text{Prob}(Px, P \rightarrow P_1 \,\&\, P_1 \rightarrow P_2 \,\&\, \ldots \,\&\, P_n \rightarrow Q \,\&\, S) =$
$\text{Prob}(Px, S)$

(3*) $\text{Prob}(Qx, P \rightarrow Q \,\&\, S) =$
$\text{Prob}(Px, S) + \text{Prob}(\sim Px, S) \times$
$\text{Prob}(Qx, \sim Px \,\&\, P \rightarrow Q \,\&\, S).$

It is clear that these are not true in general. Suppose, for example, that S is the statement that x does not have property Q. Then (1*) will be false, since $\text{Prob}(Px, P \rightarrow Q \,\&\, S)$ will have to be equal to zero, while $\text{Prob}(Px, S)$ will be non-zero. So the above relations can only hold if some restriction is placed upon the range of S.

What restriction is needed? Let us consider some other cases where the generalized postulates break down. First, a case that is very similar to the one just mentioned. Suppose that the evdience is not that x does not have property Q, but only that a large number of reliable observers claim to have seen that x does not have property Q. Then, while Prob(Px, S) might well be reasonably high, Prob(Px, $P \to Q$ & S), Prob(Px, $P \to P_1$ & $P_1 \to P_2$ & . . . & $P_n \to Q$ & S), and Prob(Qx, $P \to Q$ & S) will surely be quite low, so that (1*), (2*), and (3*) will all be false.

A different sort of case is this. Suppose that S contains the information, first, that x has property R, and secondly, that it is a statistical law that the probability that something with property R has property Q is equal to some very small value, k. Given such an S, Prob(Px, S) may be quite high, but Prob(Px, $P \to Q$ & S), Prob(Px, $P \to P_1$ & $P_1 \to P_2$ & . . . & $P_n \to Q$ & S), and Prob(Qx, $P \to Q$ & S) will all be very low, so that neither (1*) nor (2*) nor (3*) will be true.

It would seem, then, that something at least as strong as the following restriction must be imposed upon the range of S:

Restriction T: S must satisfy the following conditions:

    (1) It is not the case either that S entails that x has property Q or that S entails that x does not have property Q, and similarly with respect to properties $P_1, P_2, . . . P_n$;

    (2) It is not the case that S entails the existence of states of affairs that there is reason to believe, given S, are causally dependent either upon x's having property Q or upon x's not having property Q, and similarly with respect to properties $P_1, P_2, . . . P_n$;

    (3) It is not the case that S entails the existence of any statistical law that has implications concerning the probability that x has property Q, or concerning the probability that x has properties $P_1, P_2, . . . P_n$.

The first and third conditions seem straightforward enough. The second condition, however, calls for comment, given that the expression 'causally dependent' introduces the idea of a causal relation between states of affairs, which has not yet been explained. What would have to be done, at this point, would be to appeal to some supervenience-type analysis, in order to eliminate the reference to causal relations in favour of reference to causal laws. The latter could then be expressed by means of the $\to$ notation.

If the above restriction is adopted, we have the following postulates:

(M_1): Prob$(Px, P \rightarrow Q \,\&\, S)$ = Prob(Px, S),
provided that S satisfies requirement T;

(M_2): Prob$(Px, P \rightarrow P_1 \,\&\, P_1 \rightarrow P_2 \,\&\, \ldots \,\&\, P_n \rightarrow Q \,\&\, S)$ = Prob(Px, S),
provided that S satisfies requirement T;

(M_3): Prob$(Qx, P \rightarrow Q \,\&\, S)$ =
Prob(Px, S) + Prob$(\sim Px, S)$ $\times$
Prob$(Qx, \sim Px \,\&\, P \rightarrow Q \,\&\, S)$,
provided that S satisfies requirement T.

These three postulates seem to me to be correct. They do not, however, provide a complete basis for a theory of causation. In the first place, notice that (M_3) differs from (M_1) and (M_2) in the following, important respect. (M_1) and (M_2) specify the probability that a certain state of affairs will obtain, conditional upon the existence of a certain law or laws, in terms of probabilities that are *not* conditional upon the existence of any laws. (M_3) does not do this, since it asserts that the probability of Qx, given that $P \rightarrow Q$ and S, is related to the probability of Qx, given that $\sim Px$ and $P \rightarrow Q$ and S. So unless the latter probability can be specified in terms of probabilties that are not conditional upon the existence of causal laws, the postulates will not suffice to determine the relevant probabilities.

The crucial question, accordingly, is that of the value of Prob$(Qx, \sim Px \,\&\, P \rightarrow Q \,\&\, S)$. What I want to suggest is that this value is correctly given by:

$$\text{Prob}(Qx, \sim Px \,\&\, P \rightarrow Q \,\&\, S) = \text{Prob}(Qx, \sim Px \,\&\, S).$$

The rationale here is that the existence of the causal law $P \rightarrow Q$ can only affect the probability of some state of affairs if the situation is such that it could involve an instance of that law. If, therefore, the situation is one where $\sim Px$ is the case, and one is considering the likelihood that x will have some other property, information concerning the existence of the causal law $P \rightarrow Q$ can make no difference to the conditional probability.

One could, therefore, add the following as a fourth postulate:

(4) Prob$(Qx, \sim Px \,\&\, P \rightarrow Q \,\&\, S)$ = Prob$(Qx, \sim Px \,\&\, S)$,
provided that S satisfies requirement T.

However the underlying justification just given does not depend in any way upon its being the probability of Qx that one is considering. It applies equally well if one is considering the likelihood that x possesses some other property R. Consequently it is possible to formulate a more general, fourth postulate:

(M_4): Prob$(Rx, \sim Px \ \& \ P \rightarrow Q \ \& \ S) = $ Prob$(Rx, \sim Px \ \& \ S)$, provided that S satisfies requirement T.

Given postulates (M_3) and (M_4), one can immediately derive a statement that, as in the case of postulates (M_1) and (M_2), specifies the probability of a state of affairs, conditional upon the existence of a causal law, in terms of probabilities that are not conditional upon the existence of causal laws. Given that the statement follows from (M_3) and (M_4), it is not necessary to treat it as a postulate. However, I believe that it is best, in the present context, to set aside economy of formulation in favour of explicitness with regard to fundamental relationships. We have, therefore, the following as a fifth, non-independent postulate:

(M_5): Prob$(Qx, P \rightarrow Q \ \& \ S) = $
Prob$(Px, S) + $ Prob$(\sim Px, S) \times $ Prob$(Qx, \sim Px \ \& \ S)$, provided that S satisfies requirement T.

Are there any other postulates that are needed in order to set out a theory of causal laws? One final addition that is obviously needed is a postulate that asserts that causal laws are laws:

(M_6): If $P \rightarrow Q$, then it is a law that, for all x, if Px, then Qx.

Beyond this, however, I do not think that any additional postulates are required.

These six postulates could, in fact, be reduced to three. First, as we have seen, (M_5) is derivable from (M_3) and (M_4). Second, (M_1) follows immediately from (M_2) when n is set equal to zero. Finally, (M_3) is derivable from postulates (M_1) and (M_6). For, first, it is a truth of probability theory that:

Prob $(H, E) = $
Prob$(H, F \ \& \ E) \times $ Prob$(F, E) + $ Prob$(H, \sim F \ \& \ E) \times $ Prob$(\sim F, E)$.

Applying this to Prob$(Qx, P \rightarrow Q \ \& \ S)$ gives:

$\text{Prob}(Qx, P \rightarrow Q \& S) =$
$\text{Prob}(Qx, Px \& P \rightarrow Q \& S) \times \text{Prob}(Px, P \rightarrow Q \& S) +$
$\text{Prob}(Qx, {\sim}Px \& P \rightarrow Q \& S) \times \text{Prob}({\sim}Px, P \rightarrow Q \& S).$

But given postulate (M_6), $Px \& P \rightarrow Q$ logically entails Qx, so that $\text{Prob}(Qx, Px \& P \rightarrow Q \& S)$ must be equal to one. Also, in view of postulate (M_1), it must be the case both that $\text{Prob}(Px, P \rightarrow Q \& S)$ is equal to $\text{Prob}(Px, S)$, and that $\text{Prob}({\sim}Px, P \rightarrow Q \& S)$ is equal to $\text{Prob}({\sim}Px, S)$. When these three substitutions are made in the above, one obtains:

$\text{Prob}(Qx, P \rightarrow Q \& S) =$
$\text{Prob}(Px, S) + \text{Prob}({\sim}Px, S) \times$
$\text{Prob}(Qx, {\sim}Px \& P \rightarrow Q \& S),$

that is, postulate (M_3).

Given the above postulates, it is now a relatively straightforward matter to set out quite a simple account of the nature of causal laws. The basic strategy involved in doing so will be familiar, given the earlier discussion of the truth conditions of nomological statements in Part II. In the first place, then, basic causal laws will be identified with second-order states of affairs consisting of universals standing in a certain relation—a relation which I shall refer to as that of direct causal necessitation. Derived causal laws, in turn, can be identified with second-order states of affairs consisting of universals that are interrelated via the ancestral of the relation of direct causal necessitation. Finally, the relation of direct causal necessitation can be specified by the postulates just set out: direct causal necessitation is that unique relation between universals which satisfies those six postulates.

As the postulates stand, however, there is no reference to any relation between universals. $P \rightarrow Q$ simply says that it is a causal law that if something has property P, it also has property Q. It does not refer to any relation between those properties. So the postulates need to be altered slightly, in order to incorporate reference to a relation between universals.

Let us introduce the term '$\rightarrow_i$' as a predicate whose intended interpretation is such that the sentence '$P \rightarrow_i Q$' says that properties P and Q stand in the relation of direct causal necessitation. Replacing $P \rightarrow Q$ with $P \rightarrow_i Q$ throughout (M_1) to (M_6) will then give us the postulates that are needed:

(R_1): $\mathrm{Prob}(Px, P \to_i Q \,\&\, S) = \mathrm{Prob}(Px, S)$,
provided that S satisfies requirement T;

(R_2): $\mathrm{Prob}(Px, P \to_i P_1 \,\&\, P_1 \to_i P_2 \,\&\, \ldots \,\&\, P_n \to_i Q \,\&\, S)$
$= \mathrm{Prob}(Px, S)$,
provided that S satisfies requirement T;

(R_3): $\mathrm{Prob}(Qx, P \to_i Q \,\&\, S) =$
$\mathrm{Prob}(Px, S) + \mathrm{Prob}(\sim Px, S) \times$
$\mathrm{Prob}(Qx, \sim Px \,\&\, P \to_i Q \,\&\, S)$,
provided that S satisfies requirement T;

(R_4): $\mathrm{Prob}(Rx, \sim Px \,\&\, P \to_i Q \,\&\, S) = \mathrm{Prob}(Rx, \sim Px \,\&\, S)$,
provided that S satisfies requirement T;

(R_5): $\mathrm{Prob}(Qx, P \to_i Q \,\&\, S) =$
$\mathrm{Prob}(Px, S) + \mathrm{Prob}(\sim Px, S) \times \mathrm{Prob}(Qx, \sim Px \,\&\, S)$,
provided that S satisfies requirement T;

(R_6): If $P \to_i Q$, then it is a law that, for all x, if Px, then Qx.

A theory of causal laws, and of the relation of direct causal necessitation, can now be expressed as follows:

1. *Causal Laws.* It is a basic causal law that if something has property M, it also has property N, if and only if $M \to_i N$, that is to say, properties M and N stand in the relation of direct causal necessitation.

It is a derived causal law that if something has property M, it also has property N, if and only if properties M and N are related via the ancestral of the relation of direct causal necessitation, but not by that relation itself.

Something is a causal law if and only if it is either a basic causal law or a derived causal law.

2. *The Relation of Direct Causal Necessitation.* The relation of direct causal necessitation, $\to_i$, satisfies the following conditions:

(1) $\to_i$ is a genuine, second-order relation between universals;
(2) for any particular x, any properties $P, P_1, P_2, \ldots P_n, Q$, and R, and any statement S, $\to_i$ satisfies the following postulates:

(R_1): $\mathrm{Prob}(Px, P \to_i Q \,\&\, S) = \mathrm{Prob}(Px, S)$,
provided that S satisfies requirement T;

(R_2): $\mathrm{Prob}(Px, P \to_i P_1 \,\&\, P_1 \to_i P_2 \,\&\, \ldots \,\&\, P_n \to_i Q \,\&\, S) =$
$\mathrm{Prob}(Px, S)$,
provided that S satisfies requirement T;

(R_3): $\text{Prob}(Qx, P \rightarrow_i Q \, \& \, S) =$
$\text{Prob}(Px, S) + \text{Prob}(\sim Px, S) \times$
$\text{Prob}(Qx, \sim Px \, \& \, P \rightarrow_i Q \, \& \, S)$,
provided that S satisfies requirement T;

(R_4): $\text{Prob}(Rx, \sim Px \, \& \, P \rightarrow_i Q \, \& \, S) = \text{Prob}(Rx, \sim Px \, \& \, S)$,
provided that S satisfies requirement T;

(R_5): $\text{Prob}(Qx, P \rightarrow_i Q \, \& \, S) =$
$\text{Prob}(Px, S) + \text{Prob}(\sim Px, S) \times \text{Prob}(Qx, \sim Px \, \& \, S)$,
provided that S satisfies requirement T;

(R_6): If $P \rightarrow_i Q$, then it is a law that, for all x, if Px, then Qx.

The relation of direct causal necessitation, $\rightarrow_i$, can then be explicitly defined by applying the Ramsey/Lewis approach to theoretical terms to the theory of the relation of causal necessitation just set out. Direct causal necessitation will be that unique relation which satisfies the relevant open formula corresponding to conditions (1) and (2).

It may be helpful, at this point, to look back over the line of thought that has led to the above account. We think of causes as bringing about their effects, where this relation of bringing about is asymmetric and has, so to speak, a direction. I suggested that the way to capture these features is by means of the idea that causal relations enter into the determination of certain posterior probabilities in a way that differs from that of mere nomological relations.

On the supervenience view of causal relations, however, it is not causal relations but causal laws that are fundamental. Consequently, in order to formulate a supervenience account, what are needed are postulates that specify how causal laws enter into the determination of probabilities. Thus one was led to the formulation of postulates (M_1) to (M_6), which appear to capture the relevant asymmetry with respect to the determination of probabilities.

The next step involved the introduction of the view of laws defended earlier. This led to the replacement of postulates (M_1) to (M_6) by postulates incorporating explicit reference to a second-order relation between universals—namely, postulates (R_1) to (R_6). These latter postulates were then used to set out a theory of the relation of direct causal necessitation, and hence of causal laws. Finally, given such a theory, explicit definition of the relation of direct causal necessitation is simply a matter of applying the general Ramsey/Lewis approach to the meaning of theoretical terms.

Some brief comments are now in order. In the first place, the above account is based upon a supervenience view of causal relations. Even if such a view turns out to be correct, however, the above account needs some modification. The reason will be apparent, given the discussion of nomological relations in Chapter 2. Namely, it would seem to be logically possible for it to be a causal law, for example, that if something has property P, then there will be something else to which it stands in relation R, and which *lacks* property S. Can such a law be identified with a second-order state of affairs involving two universals standing in the relation of direct causal necessitation? If there are no negative universals, the answer would seem to be no. For if there need not be any universal that is present in everything that lacks property S, then neither need there be any universal that is present in everything that stands in relation R to something that lacks property S. Some relation other than that of direct causal necessitation will therefore have to be introduced to handle this sort of case.

What one wants, of course, is a completely general account of what it is for something to be a second-order relation of such a sort that its holding among universals constitutes a causal law. But here, as in the case of probabilistic laws, it will not really be necessary to set out such a general account, since the lines along which such an analysis would proceed are parallel to those followed in the treatment of non-probabilistic laws in Chapter 2.

Secondly, I have spoken above of *the* relation of direct causal necessitation. This implies that there is only one relation that satisfies postulates (R_1) to (R_6). I believe that this is the case. However, it is not really possible to demonstrate that this is so, in the absence of some theory concerning the individuation of relations. But, as I shall be indicating in section 8.4, it can be shown that nothing untoward follows if it turns out that there is more than one relation between universals that satisfies postulates (R_1) to (R_6), with the result that direct causal necessitation is, strictly speaking, not a relation, but a quasi-relation.

Finally, although I have attempted to make clear the basic line of thought underlying the above account of causal laws, I have not offered any detailed defence. This will be done later—primarily in sections 8.4 and 9.2, and in the Appendix. In section 8.4, and in the Appendix, I shall argue that the above account is formally sound, in the sense that it follows from the analysis that causal laws have those formal properties that they are normally taken to have. In section 9.2, I shall argue that the above account of causal laws is epistemologically satisfactory.

8.3.3 *Version Two: Causally Bringing About as Basic*

In this section, I shall consider whether the intuitive idea, that causal facts are theoretical ones that determine the direction of the logical transmission of probabilities, can provide an analysis of causal concepts if a singularist view is adopted. In that case, the fundamental causal concept is not that of a causal law, but that of the relation that obtains between two states of affairs when the one brings about the other. The question, therefore, is whether it is possible to formulate postulates which specify how such causal relations between states of affairs enter into the determination of probabilities.

In attempting to set out satisfactory postulates, it will be convenient to have a brief way of representing the fact that one state of affairs causally brings about another. I shall employ the following symbolism:

$$Pa \rightarrow Qa.$$

The intended interpretation of this is that the state of affairs consisting of a's having property P causally brings about the state of affairs consisting of a's having property Q.

Let us now turn to the problem of setting out the desired postulates. A natural idea, given the account of causal laws advanced in the preceding section, is to try to modify the postulates arrived at there. Consider, then, postulate (R_1). The most straightforward modification involves replacing $P \rightarrow_i Q$ with $Px \rightarrow Qx$. This gives one:

$(R_1{}^*)$: $\mathrm{Prob}(Px, Px \rightarrow Qx \ \& \ S) = \mathrm{Prob}(Px, S)$,
  provided that S satisfies requirement T.

Is postulate $(R_1{}^*)$ acceptable? The answer is that it is not. Since the state of affairs which is x's having property P must exist if it is to give rise to some other state of affairs, $\text{Prob}(Px, Px \rightarrow Qx \,\&\, S)$ must be equal to one. $\text{Prob}(Px, S)$, in contrast, need not be equal to one. So $(R_1{}^*)$ is simply false.

Can anything be done about this problem? Consider, once again, the case of the coin and the die. This time, however, we need to suppose that while there is no law at all connecting the coin's coming up heads with the die's coming up six, none the less, it is sometimes the case that the coin's coming up heads causally brings it about that the die comes up six. The question then is how, given this singularist situation, one can make sense of the idea of a logical transmission of probabilities from cause to effect.

The stumbling block is the fact that, while the prior probability of the coin's coming up heads is very different from that of the die's coming up six, their posterior probabilities, relative to the fact that the coin's coming up heads causes the die to come up six, are both equal to one, so that there is no interesting relation between the prior and the posterior probabilities. Therefore, if the idea of a transmission of probabilities from cause to effect is to gain a foothold, the relevant posterior probabilities cannot be ones that are conditional upon the fact that the coin's coming up heads causally brings it about that the die comes up six.

The source of the problem is that the proposition that the coin's coming up heads causes the die to come up six, since it implies that the coin does come up heads and that the die does come up six, provides too much information. What is needed is information about the causal relation which does not specify whether the coin comes up heads or whether the die comes up six, and consequently does not necessarily result in posterior probabilities that are equal to one.

There is a standard ploy that sometimes generates acceptable solutions to problems of this general sort. If some statement, C, contains information of a desired sort, but also has some undesired entailment, D, one can try replacing C with the disjunctive statement, Either C or not D. This statement will not entail D, and it may still provide the information needed.

One way of applying this technique in the present case is by replacing $Px \rightarrow Qx$ by $Px \rightarrow Qx \,\text{v}\, {\sim}Px$. When this substitution is made in $(R_1{}^*)$, one has the following postulate:

(U_1): $\text{Prob}(Px, (Px \rightarrow Qx \,\text{v}\, {\sim}Px) \,\&\, S) = \text{Prob}(Px, S)$
   provided that S satisfies requirement T.

Unfortunately, it can be seen that U_1 is also unacceptable. For, given that S is logically equivalent to ((Px & $Px \rightarrow Qx$) v (Px & $\sim(Px \rightarrow Qx)$)) v $\sim Px$) & S, we have, by the substitution of logically equivalent expressions, that Prob(Px, S) is equal to Prob(Px, ((Px & $Px \rightarrow Qx$) v (Px & $\sim(Px \rightarrow Qx)$)) v $\sim Px$) & S). Similarly, in view of the fact that $Px \rightarrow Qx$ is analytically equivalent to Px & $Px \rightarrow Qx$, we also have that Prob(Px, ($Px \rightarrow Qx$ v $\sim Px$) & S) is equal to Prob(Px, ((Px & $Px \rightarrow Qx$) v $\sim Px$) & S). Comparing these two, one can see that Prob(Px, S) will be greater than Prob(Px, ($Px \rightarrow Qx$ v $\sim Px$) & S) unless Prob(Px & $\sim(Px \rightarrow Qx)$, S) is equal to zero. Since the latter will not be true in general, (U_1) must be rejected.

A thought that one might have at this point is that perhaps the unsoundness of (U_1) results from the fact that, in employing the statement, $Px \rightarrow Qx$ v $\sim Px$, one is using, in effect, a material conditional, where what is needed is the indicative conditional: If Px, then $Px \rightarrow Qx$. What would happen if one used indicative conditionals in formulating the postulates?

In the case of the first postulate, (U_1) would be replaced by:

(U_1^*): Prob(Px, (If Px, then $Px \rightarrow Qx$) & S) = Prob(Px, S).

This does seem to escape the sort of objection just raised against (U_1). However, I believe that there is still a problem, and one which emerges once one asks what the basis of the relevant indicative conditional could be. On the one hand, it cannot rest simply upon the corresponding material conditional, for then one would be back to the difficulty to which (U_1) was exposed. But, on the other hand, if it rests upon a causal law connecting properties P and Q, then the analysis of the singularist causal connection between states of affairs—a connection that can exist in the absence of any underlying law—would seem to presuppose the existence of underlying laws. The problem, therefore, is to find some unpacking of the relevant indicative conditional that is neither too strong nor too weak. The prospects do not seem especially bright.

There is, in short, a serious problem of finding acceptable postulates that will enable one to analyse the singularist conception of causation in terms of the logical transmission of probabilities. Moreover, even if one succeeded in finding postulates that had no objectionable entailments, one would still have to overcome the traditional difficulty faced by any singularist account—namely, that of showing how, if causal relations do not presuppose underlying laws, the relation of causation between states of affairs differs from the non-causal relation involved when one state of affairs merely happens to *accompany* another.

If there were a powerful argument in support of a singularist conception, it would be reasonable to suppose that a satisfactory answer must be available. The situation is quite different, however, if, as I believe, the most powerful support that can be offered for a singularist conception is along the lines of the arguments set out in Chapter 6. For those arguments are arguments against the supervenience view, and, as such, leave completely open the choice between the singularist account and the intermediate view. Accordingly, if it turns out that the intermediate approach escapes the difficulties associated with the singularist account, there will be no reason to struggle further with the problem of trying to make sense of the singularist conception of causation.

8.3.4 *Version Three: The Intermediate View*

The intermediate account of causal relations shares with the supervenience approach the contention that states of affairs cannot be causally related unless they fall under some causal law, but it rejects the claim that causal laws, together with the non-causal properties of, and relations among things, serve to determine what events are causally related. But how is one to make sense of this ontologically? The answer, at least given the present view of laws, would seem to be that when two states of affairs are causally connected, there must be two relations involved—the second-order relation between the universals that enter into the underlying causal law, along with some special first-order relation between the two states of affairs. If this is right, then one of the fundamental questions to be answered concerns how those two relations are connected.

In tackling this issue, I think it is important to realize that there are two rather different ways in which one might view an intermediate account. One might view it as a supplemented singularist conception, or as a supplemented supervenience account, and this will have a bearing upon which of the two special relations is more basic. I shall treat it as a supplemented supervenience account. My first reason for doing so will be obvious. Given that I have been unable to develop a set of postulates which successfully capture a singularist conception of causation, there is no possibility of setting out a supplemented version of such an account. But secondly, it seems to me that if one could make sense of a singularist view, its greater simplicity would render it preferable to the corresponding, intermediate view. A viable intermediate account, therefore, will necessarily be a supplemented version of a supervenience account.

My approach, therefore, will be to model an intermediate account upon the supervenience version set out above. The idea will be to preserve, as far as possible, the postulates that characterize the supervenience approach, but to strengthen them in order to introduce the additional, first-order relation between states of affairs that is required for an intermediate account.

Before doing this, however, we need to consider what modifications are needed in the notation being employed. One thing that will certainly be required is a way of representing the special, first-order relation between states of affairs. To do so, let us use

$CaPbQ$

where this expression is to be interpreted as follows. First, there is the case where properties P and Q are intrinsic properties of individuals. Then, regardless of whether a and b are distinct individuals or not, the expression '$CaPbQ$' will be interpreted as meaning that the state of affairs which consists of a's having property P stands in relation C to the state of affairs which consists of b's having property Q, where C is such that, if $CaPbQ$, and there is an appropriate causal law, then a's having property P causes b to have property Q. Second, there is the case where a and b are one and the same individual, and where property Q, rather than being an intrinsic property of an object, is a relational property. When this is so, to say that a has property Q will be to say, for the appropriate relation R, and the appropriate intrinsic property I, that there is some individual x such that x has property I and stands in relation R to a. In such a case, one does not want to say that a's having property P stands in relation C to the state of affairs which consists of a's having property Q. Relation C obtains, rather, between a's having property P, and the state of affairs which consists of x's having property I, where x is the unique individual which stands in relation R to a, and which also possesses property I. So this is what the expression '$CaPbQ$' must be taken as asserting when Q is a relational property, and when a and b are not distinct individuals. This second case is important, of course, because of the notational convention adopted earlier—in section 8.3.2—to simplify the representation of causal laws.

What it is for there to be a causal law will be slightly different on the intermediate model than on the supervenience one, since, on the former view, but not on the latter, the existence of causal laws entails the existence of the special first-order relation C. It will be convenient, however, to retain the same symbolism, i.e. $P \rightarrow Q$, interpreted as expressing the proposition that it is a causal law that if something has property P, then it has property Q.

It was noted earlier that the fact that this way of expressing causal laws is in terms of properties, P and Q, possessed by one and the same thing, is not really a limitation, since Q can be a relational property. Thus, if 'x has property Q' means the same as 'There is a y that stands in relation R to x, and which has property I', then $P \rightarrow Q$ will express the proposition that it is a causal law that if x has property P, then there is some y, standing in relation R to it, which has property I.

The same point applies here, the only difference being that the scope of acceptable relational properties is, it would seem, slightly wider in the context of an intermediate approach to causation. For consider the relational property which x has when there is a y, with property I, and which is distinct from x. Since there may very well be a number of things with property I and distinct from x, if relational properties of this sort could enter into causal laws as conceived of on the supervenience model, there would be nothing that made it the case that one state of affairs involving an instance of property I, rather than some other, was causally related to x's having property P. In contrast, if an intermediate approach is adopted, the causally related states of affairs will be connected by the special, first-order relation, C. In short, on the supervenience approach, the relation involved in the relational property, Q, needs to be one that, relative to x, picks out a unique object, whereas on the intermediate approach, this need not be the case. Relational properties of the broader sort mentioned above are perfectly satisfactory.

We can now turn to the question of the appropriate postulates. Recall that, although the supervenience view was set out in terms of six postulates, three of those are not really necessary as postulates, since they can be derived from the others. As a consequence, attention can be confined to postulates (M_2), (M_4), and (M_6):

(M_2): $\text{Prob}(Px, P \to P_1 \,\&\, P_1 \to P_2 \,\&\, \ldots \,\&\, P_n \to Q \,\&\, S) = \text{Prob}(Px, S)$;

(M_4): $\text{Prob}(Rx, \sim Px \,\&\, P \to Q \,\&\, S) = \text{Prob}(Rx, \sim Px \,\&\, S)$;

(M_6): If $P \to Q$, then it is a law that, for all x, if Px, then Qx.

The intuition motivating (M_2) was that the probability that something will have property P should not be affected by information concerning the other sorts of states of affairs that would thereby be brought about, provided that S is subject to an appropriate restriction T. As this seems no less plausible on an intermediate view than on the supervenience approach, (M_2) certainly seems acceptable.

The idea underlying (M_4) was that the fact that the possession of property P causally brings about the possession of some other property Q should not affect the likelihood that x has property R, provided that it is a part of the evidence that x does not have property P: causal laws only affect probabilities in situations where their application is not precluded. Since the shift from a supervenience approach to an intermediate one in no way alters the plausibility of this claim, (M_4) would also seem to be sound.

Finally, (M_6): that (M_6) is true, given the intermediate approach, is clear. But it is also apparent that it is not sufficiently strong. The assertion that it is a causal law that the possession of property P brings about the possession of property Q must imply, on an intermediate approach, that properties P and Q are nomologically connected not only with each other, but also with the special first-order relation, C. What one needs, then, is something like:

$(M_6{}^*)$: If $P \to Q$, then it is a law that, for all x, if Px, then Qx and $CxPxQ$.

With this slight strengthening of (M_6), I believe that one has all that is needed for the formulation of an intermediate account of causation. The result will be an account that is very close to the supervenience one. For, as before, (M_1) is simply a special case of

(M_2), and, since (M_6^*) entails (M_6), (M_3) is derivable from (M_1) together with (M_6^*). Finally, (M_5) follows from (M_3) and (M_4). The two accounts differ, therefore, only with respect to one postulate.

Given postulates (M_1) to (M_5), plus (M_6^*), the strategy is the same as in the case of the supervenience account—namely, the postulates in question are modified slightly, so as to reflect the theory of laws set out in Part II, according to which laws involve second-order relations among universals. The result is that basic causal laws are identified with laws in which the second-order relation is that of direct causal necessitation, and where the nature of that relation is explained in terms of the postulates in question.

Postulates (M_1) to (M_5), plus (M_6^*), need to be modified, therefore, to incorporate reference to the relation of direct causal necessitation. As before, this is simply a matter of replacing all occurrences of expressions such as $P \rightarrow Q$ with expressions of the form $P \rightarrow_i Q$, where this is interpreted as meaning that the universals, P and Q, stand in the second-order relation of direct causal necessitation.

The resulting postulates, S_1 to S_6, can then be used to formulate the following, intermediate theory of causation:

> The relation of direct causal necessitation, $\rightarrow_i$, and the relation of causation, C, satisfy the following conditions:
>
> (1) $\rightarrow_i$ is a genuine, second-order relation between universals;
>
> (2) C is a first-order relation between states of affairs;
>
> (3) for any particular x, any properties P, P_1, P_2, . . . P_n, Q, and R, and any statement S, $\rightarrow_i$ satisfies the following postulates:
>
> (S_1): $\text{Prob}(Px, P \rightarrow_i Q \& S) = \text{Prob}(Px, S)$,
> provided that S satisfies requirement T;
>
> (S_2): $\text{Prob}(Px, P \rightarrow_i P_1 \& P_1 \rightarrow_i P_2 \& \ldots \& P_n \rightarrow_i Q \& S)$
> $= \text{Prob}(Px, S)$,
> provided that S satisfies requirement T;
>
> (S_3): $\text{Prob}(Qx, P \rightarrow_i Q \& S) =$
> $\text{Prob}(Px, S) + \text{Prob}(\sim Px, S) \times$
> $\text{Prob}(Qx, \sim Px \& P \rightarrow_i Q \& S)$,
> provided that S satisfies requirement T;

(S_4): Prob(Rx, $\sim Px$ & $P\to_i Q$ & S) = Prob(Rx, $\sim Px$ & S),
provided that S satisfies requirement T;

(S_5): Prob(Qx, $P\to_i Q$ & S) =
Prob(Px, S) + Prob($\sim Px$, S) × Prob(Qx, $\sim Px$ & S),
provided that S satisfies requirement T;

(S_6): If $P\to_i Q$, then it is a law that, for all x, if Px, then Qx
and $CxPxQ$,
provided that S satisfies requirement T.

This theory contains, of course, two terms whose meaning has not yet been defined—namely, $\to_i$ and C. But once again, the idea is simply to apply the Ramsey/Lewis approach to the meaning of theoretical terms. Thus, the two undefined terms can be replaced by variables, thereby transforming the theory into an open statement. One can then explain the meanings of the terms $\to_i$ and C in terms of the first and second elements, respectively, of the unique ordered pair of universals that satisfies the open formula in question.

There is one final point that needs to be discussed, and that is the difficulty, for any intermediate account of causation, which was advanced in Chapter 6. The problem, essentially, was that if causation involves *both* laws and some first-order relation between states of affairs, then it would seem that it should be possible for there to be states of affairs which stand in the relation in question, even if there is no relevant law. But if this is possible, how does the intermediate view differ from the singularist, other than in an arbitrary decision to interpret causal language in such a way that it is applicable only when the states of affairs that stand in the special first-order relation also happen to fall under some causal law?

The answer to this objection is as follows. If a singularist view of causation were tenable, there could be worlds which contained causally related events, even though there were no causal laws at all. Consider, then, a world without causal laws, but where there are a, b, P, and Q, such that $CaPbQ$. Is it merely an arbitrary verbal stipulation to refuse to say that, in that world, a's having property P would be causally related to b's having property Q? Not in the light of the discussion in section 8.3.3. For what emerged there is that it does not seem possible to make sense of a singularist view in terms of the idea of a logical transmission of probabilities. But if that is so, and if it is also the case, as I am maintaining, that the idea of a logical transmission of probabilities lies at the very heart of the concept of causation, then we are forced to conclude that, in a world where there were *a*, *b*, *P*, and *Q* such that *CaPbQ*, but no causal laws, it would not be the case that *a*'s having property *P* was causally related to *b*'s having property *Q*.

The point, in short, is that, on the present intermediate view, any relation, *C*, which is the special, first-order relation that obtains between any two causally related states of affairs, has that status *not by virtue of its intrinsic nature*, but by virtue of the fact that it is the relation which enters into all causal laws in a certain way—and where causal laws are laws which obtain by virtue of a second-order relation between universals which determines the direction of the logical transmission of probabilities. The primary locus of causal fact, on the intermediate model advanced here, is the second-order relation which enters into causal laws. The first-order relation between states of affairs, in contrast, plays a role in causal facts *only* because of its connection with the basic, second-order relation between universals.

8.4 FORMAL DEVELOPMENT: I. THE SUPERVENIENCE VIEW

In the preceding sections, I have attempted both to motivate, and to render plausible, certain analyses of fundamental causal concepts, by appealing to what seems to me to be a very important intuition concerning causation—the idea, namely, that causal facts are ones that determine the direction of the logical transmission of probabilities. It is, I suggest, a strong argument in favour of those analyses that they enable one to capture that idea, and to express it in a precise fashion.

In the discussion that follows, I shall attempt to develop another central line of argument in support of the above analyses, by indicating how it is possible to demonstrate, given the above accounts, that causal laws and causal relations possess precisely those formal properties that they are normally taken to possess. In doing so, I shall offer a reasonably detailed sketch of the main steps involved in showing that the crucial formal properties in question are derivable from the accounts offered above. Two rather long, formal arguments have, however, been placed in an

Appendix, and some arguments which were especially straightforward have been omitted.

The discussion in this section will be confined to the supervenience approach, and it is divided into four sections. The first deals with some important formal properties of the relation of direct causal necessitation; the second, with the ancestral of that relation; the third, with some further properties of direct causal necessitation; and the fourth, with causal laws, causally necessary and causally sufficient conditions, and causal priority.

8.4.1 *The Relation of Direct Causal Necessitation*

What are some of the more important formal properties of direct causal necessitation? Since it is the relation involved in *basic* causal laws, it need not be transitive. It must, however, be both asymmetric and irreflexive. It must also be the case that contraposition can never obtain, and that causal loops, of any sort, are impossible. The main task here, therefore, will be to show that direct causal necessitation possesses those properties.

Before turning to a consideration of the central formal properties of the relation of direct causal necessitation, however, there is a preliminary result that is needed. For recall that, while the initial idea was to define direct causal necessitation as the relation among universals which satisfies postulates (R_1) to (R_6), the problem arises that there is no obvious guarantee that only one relation between universals satisfies those postulates. If it is true, then, that there can be only one genuine relation between universals which satisfies postulates (R_1) to (R_6), some proof of this is needed. The difficulty is that it seems clear that such a proof will turn upon some account of the individuation of universals, and at present I am unable to offer such an account. I propose to follow, therefore, the next best route. Assume, for the sake of argument, that there may be more than one genuine relation between universals which satisfies those postulates. Direct causal necessitation will then have to be defined as the *union* of all of those relations. What then needs to be shown is that when direct causal necessitation is thus defined, it satisfies the postulates in question. Consider, then, the following six theorems:

Theorem 1: *Direct Causal Necessitation Satisfies Postulate (R_2).*

Theorem 2: *Direct Causal Necessitation Satisfies Postulate* (R_1).
Theorem 3: *Direct Causal Necessitation Satisfies Postulate* (R_6).
Theorem 4: *Direct Causal Necessitation Satisfies Postulate* (R_3).
Theorem 5: *Direct Causal Necessitation Satisfies Postulate* (R_4).
Theorem 6: *Direct Causal Necessitation Satisfies Postulate* (R_5).

In order to show that the quasi-relation of direct causal
necessitation, defined as the union of all genuine relations between
universals satisfying postulates (R_1) to (R_6), satisfies any particular
postulate, it suffices to show that the union of any *two* relations or
quasi-relations that satisfy all the postulates must also satisfy the
postulate in question.

Theorems 1 and 5 involve only the straightforward application
of some elementary theorems of probability theory. The arguments
are, however, rather tedious, especially in the case of Theorem 1,
and for that reason it seems best to omit them.

The proof of Theorem 3, on the other hand, can be stated very
briefly. Suppose that $\rightarrow_i$ and $\rightarrow_j$ are two relations, or quasi-
relations, satisfying postulates (R_1) to (R_6), and let $\rightarrow_k$ be the
union of them, i.e., $P \rightarrow_k Q$ if and only if either $P \rightarrow_i Q$ or $P \rightarrow_j Q$.
If $P \rightarrow_i Q$, it follows by virtue of the fact that $\rightarrow_i$ satisfies postulate
(R_6) that it is a law that, for all x, if Px, then Qx; similarly, for
$P \rightarrow_j Q$. So if $P \rightarrow_k Q$, it must be a law that, for all x, if Px, then
Qx. Any disjunctive quasi-relation of two relations (or quasi-
relations) satisfying postulates (R_1) to (R_6) must, therefore, also
satisfy postulate (R_6), and from this it follows that direct causal
necessitation must satisfy postulate (R_6).

The proofs of Theorems 2, 4, and 6 are even briefer. For given
that (R_1) is just a special case of (R_2), Theorem 2 is an immediate
corollary of Theorem 1. Similarly, given that (R_3) is derivable
from (R_1) in conjunction with (R_6), Theorem 4 is an immediate
corollary of Theorems 2 and 3. Finally, since (R_5) is derivable from
(R_3) and (R_4), Theorem 6 is an immediate corollary of Theorems 4
and 5.

So direct causal necessitation, even if it turns out to be a quasi-
relation, rather than a relation, satsifies postulates (R_1) to (R_6).
Given this preliminary result, we can now turn to the main task—
namely,that of establishing that direct causal necessitation possesses certain crucial properties. In particular, it will need to be
shown that direct causal necessitation is irreflexive and asymmetric,

that contraposition is impossible, that causal loops, of even the most general sort, are precluded, and that direct causal necessitation is related in a certain way to increase in probability.

Theorem 7: Direct Causal Necessitation is Related to Increase in Probability. An important issue concerns the relation between causation and increase in probability. With regard to this, some philosophers have suggested that events of type P cannot cause events of type Q unless events of type Q are more likely given events of type P than in their absence. We saw in section 7.5, however, that this view is untenable. None the less, there is a relation between causation and increase in probability, and I believe that the present theorem provides the correct account of it.

Compare the probability that something has property Q, given that S, with the corresponding probability given S together with the fact that it is a causal law that the possession of property P brings about the possession of property Q. As regards the former, it is a truth of probability theory that

$$\text{Prob}(Qx, S) = \text{Prob}(Qx, Px \& S) \times (\text{Prob}(Px, S) + \text{Prob}(Qx, \sim Px \& S) \times \text{Prob}(\sim Px, S).$$

As regards the latter, postulate R_5 asserts that

$$\text{Prob}(Qx, P \rightarrow_k Q \& S) = \text{Prob}(Px, S) + \text{Prob}(Qx, \sim Px \& S) \times \text{Prob}(\sim Px, S),$$

provided that S satisfies requirement T.

Comparing these two equations, one can see that, if S satisfies requirement T,

(1) $\text{Prob}(Qx, P \rightarrow_k Q \& S)$ will always be equal to, or greater than, $\text{Prob}(Qx, S)$,
(2) those two probabilities will be equal only if either
   (a) $\text{Prob}(Qx, Px \& S)$ is equal to one, or (b) $\text{Prob}(Px, S)$ is equal to zero.

This account of the relation between causation and increase in probability, in contrast to other accounts, is not, it seems, exposed to counterexamples. In addition, there are other advantages. First, the relationship, rather than being treated as an unexplained postulate, can be derived from a more basic postulate. Second, the derivation makes it clear that the relation, rather than being a completely general one, only holds if certain conditions are satisfied. Finally, given an account of probabilistic causal laws, one can show that the relation in question does *not* extend into that domain.

Theorem 8: Direct Causal Necessitation is Necessarily Irreflexive. That direct causal necessitation is necessarily irreflexive would seem to be an important fact. But how is it to be established? There are, I think, two possible lines of argument. The first is that, while a property's standing in the relation (or quasi-relation) of direct causal necessitation to itself is not precluded by postulates (R_1) to (R_5), it is ruled out by postulate (R_6), provided that one is granted the claim that a generalization which expresses a law must be only contingently true. For if there were some property, P, such that $P \rightarrow_i P$, postulate (R_6) would entail that it is a law that, for all x, if Px, then Px.

There is a second line of argument that is, I suspect, really more basic. It involves the claim that it is logically impossible for anything to stand in a genuine relation to itself.[3] The argument is that while direct causal necessitation may be only a quasi-relation, since it is a union of genuine relations, it cannot be the case that some property stands in the quasi-relation of direct causal necessitation to itself unless there is some genuine relation, satisfying postulates (R_1) to (R_6), in which that property stands to itself. If, therefore, nothing can stand in any genuine relation to itself, direct causal necessitation, even if it is only a quasi-relation, is necessarily irreflexive.

Theorem 9: Direct Causal Necessitation is Necessarily Antisymmetric. One of the most important facts about direct causal necessitation is that it is necessarily asymmetric. It seems, however, that the correct way to establish this conclusion involves first showing that it is necessarily antisymmetric—where a relation, R, is *antisymmetric* if and only if it is the case that, for any x and y, if xRy and yRx, then it necessarily follows that x is identical to y. Once it has been shown that direct causal necessitation is necessarily antisymmetric, this together with the fact that it is necessarily irreflexive will entail that it is necessarily asymmetric.

The form of argument here will be to assume that direct causal

[3] For a discussion and defence of this claim, see Armstrong's *Universals and Scientific Realism*, ii, pp. 91–3.

necessitation is not antisymmetric, and then to show that that leads to a consequence that is necessarily false. Let us assume, therefore, that there are distinct properties P and Q such that it is true both that $P \rightarrow_k Q$ and $Q \rightarrow_k P$, where $\rightarrow_k$ is the relation (or quasi-relation) of direct causal necessitation. We now need to consider whether it must be possible to find some statement S, satisfying requirement T, such that the following are true:

(1) $\mathrm{Prob}(Qx, Px \;\&\; S) \neq 1$;
(2) $\mathrm{Prob}(Px, S) \neq 0$.

The first condition seems unproblematic. On the other hand, some doubt might be raised concerning the satisfiability of the second condition. I suggest, however, that when it is kept in mind that probability values are not confined to the standard real numbers in the interval from zero to one, so that $\mathrm{Prob}(Px, S)$ could be an infinitesimal, it becomes clear that there is no problem about choosing an appropriate S which will make both (1) and (2) true.

It should be noted, however, that it would not be possible to find an S satisfying (1) if P and Q were not distinct properties. It is for this reason that the proof that follows establishes that direct causal necessitation is antisymmetric, but not that it is asymmetric.

The argument now proceeds as follows. Given an S such that (1) and (2) obtain, we have, by virtue of Theorem 7, that

(3) $\mathrm{Prob}(Qx, P \rightarrow_k Q \text{ and } S)$ is greater than $\mathrm{Prob}(Qx, S)$.

Next, by postulate (R_1) we have that

$\mathrm{Prob}(Qx, Q \rightarrow_k P \;\&\; P \rightarrow_k Q \;\&\; S) = \mathrm{Prob}(Qx, P \rightarrow_k Q \;\&\; S)$,

while by postulate (R_2) we have that

$\mathrm{Prob}(Qx, Q \rightarrow_k P \;\&\; P \rightarrow_k Q \;\&\; S) = \mathrm{Prob}(Qx, S)$.

When these are combined, we have that

(4) $\mathrm{Prob}(Qx, P \rightarrow_k Q \;\&\; S) = \mathrm{Prob}(Qx, S)$.

But statements (3) and (4) are incompatible. It is therefore impossible to have both $P \rightarrow_k Q$ and $Q \rightarrow_k P$, if P and Q are distinct properties. Direct causal necessitation is necessarily asymmetric.

Theorem 10: Direct Causal Necessitation is Necessarily Asym-

metric. This follows immediately from the fact that direct causal necessitation is necessarily both irreflexive and antisymmetric.

To sum up, we have seen that it follows from the account of direct causal necessitation offered in section 8.3.2 that it satisfies postulates (R_1) to (R_6), that it is related to increase in probability, and that it is irreflexive and asymmetric. It also needs to be shown that contraposition always fails, and that causal loops involving the relation of direct causal necessitation are impossible. This, however, is best left until section 8.4.3, since the simplest way of establishing the latter facts is by proving the corresponding theorems in the case of the ancestral of direct causal necessitation.

8.4.2 *The Ancestral of Direct Causal Necessitation*

Causal laws may be either basic or derived. When two universals stand in the relation of direct causal necessitation, one has a basic causal law. When they stand in the ancestral of that relation, one has a derived causal law.

This section deals with the properties of the ancestral of direct causal necessitation. We shall see that they are very similar to those of direct causal necessitation. The main difference is that while the ancestral is, of course, transitive, direct causal necessitation is not.

Let us use the term '$\to^*$' as a predicate whose intended interpretation is such that the sentence '$P \to^* Q$' says that property P is related to property Q via the ancestral of direct causal necessitation.

The first thing that needs to be shown is that the ancestral satisfies postulates (R_1^*) to (R_3^*), and (R_6^*)—where (R_1^*) is the postulate that results from (R_1) when every occurrence of $\to_k$ is replaced by an occurrence of $\to^*$, and similarly for the other postulates.

Theorem 11: *The Ancestral Satisfies Postulate* (R_2^*). That the ancestral of direct causal necessitation satisfies postulate (R_2^*) follows immediately from the fact that direct causal necessitation satisfies postulate (R_2).

Theorem 12: *The Ancestral Satisfies Postulate* (R_1^*). Since (R_1^*) is just a special case of (R_2^*), this is an immediate corollary of the preceding theorem.

Theorem 13: *The Ancestral Satisfies Postulate* $(R_6{}^*)$. Suppose that $P \rightarrow^* Q$. Then there must exist properties $P_1, P_2, \ldots P_n$, such that $P \rightarrow_k P_1$, $P_1 \rightarrow_k P_2$, $\ldots$ and $P_n \rightarrow_k Q$, where $\rightarrow_k$ is the relation of direct causal necessitation. By virtue of the fact that direct causal necessitation satsifies postulate (R_6), it follows that it must be a law that, for all x, if Px, then P_1x; it must be a law that, for all x, if P_1x, then P_2x; $\ldots$ and it must be a law that, for all x, if P_nx, then Qx. It therefore follows that it must be a law that, for all x, if Px, then Qx. This shows that the ancestral must satisfy $(R_6{}^*)$.

Theorem 14: *The Ancestral Satisfies Postulate* $(R_3{}^*)$. $(R_3{}^*)$ is derivable from $(R_1{}^*)$ and $(R_6{}^*)$ in just the way that (R_3) is derivable from (R_1) and (R_6). Accordingly, the fact that the ancestral satisfies $(R_3{}^*)$ follows from Theorems 12 and 13.

Theorem 15: *Causal Necessitation is Related to Increase in Probability*. In the previous section it was shown, in Theorem 7, that $\mathrm{Prob}(Qx, P \rightarrow_k Q \ \& \ S)$ will always be greater than $\mathrm{Prob}(Qx, S)$, provided that S satisfies requirement T, except when either $\mathrm{Prob}(Qx, Px \ \& \ S)$ is equal to one, or $\mathrm{Prob}(Px, S)$ is equal to zero, in which case they will be equal. A closely related result can be established in the case of the ancestral—namely, one can show that $\mathrm{Prob}(Qx, P \rightarrow^* Q \ \& \ S)$ will be greater than $\mathrm{Prob}(Qx, S)$, except for some of the cases where $\mathrm{Prob}(Qx, \sim Px \ \& \ S)$ is equal to one, or $\mathrm{Prob}(Px, S)$ is zero, when they may be equal. This result can be established in quite a straightforward manner by mathematical induction.

Readers interested in the details of the argument will find a statement of the proof, along with a more precise formulation of the theorem, in the Appendix.

Next, it needs to be shown that the ancestral of direct causal necessitation is irreflexive and asymmetric. As in the case of direct causal necessitation, to prove that the relation is asymmetric involves first proving that it is antisymmetric. The relevant arguments are as follows.

Theorem 16: *The Ancestral is Necessarily Antisymmetric*. The proof, in Theorem 9, that direct causal necessitation is necessarily antisymmetric, appealed only to postulates (R_1) and (R_2), plus Theorem 7, which concerned the relation between causation and increase in probability. As Theorems 11, 12, and 15 show that comparable propositions are true of the ancestral, the method of proof used in Theorem 9 can be paralleled to show that the ancestral of direct causal necessitation must also be necessarily antisymmetric.

Theorem 17: *The Ancestral is Necessarily Irreflexive.* Suppose that there is some property , P, such that $P \rightarrow^* P$. Then by virtue of the definition of the ancestral, it must either be the case that $P \rightarrow_k P$, or else there must be properties $P_1, P_2, \ldots P_n$ such that $P \rightarrow_k P_1$ & $P_1 \rightarrow_k P_2$ & $\ldots P_n \rightarrow_k P$. The first is ruled out by the fact that direct causal necessitation is necessarily irreflexive. On the other hand, if the second possibility were the case, then it would be true that $P \rightarrow^* P_1$ & $P_1 \rightarrow^* P$, and therefore it would be false that the ancestral of direct causal necessitation is necessarily antisymmetric. The supposition that there is some property that is related to itself via the ancestral of the relation (or quasi-relation) of direct causal necessitation therefore leads to a contradiction. So the ancestral of direct causal necessitation must be irreflexive.

Theorem 18: *The Ancestral is Necessarily Asymmetric.* Given that the ancestral of the relation (or quasi-relation) of direct causal necessitation is necessarily both antisymmetric and irreflexive, it follows immediately that it is necessarily asymmetric.

Theorem 19: *Causal Loops are Impossible.* A crucial property of the ancestral of the relation of direct causal necessitation is that contraposition is impossible—that is to say, it cannot be the case for any properties P and Q, that both $P \rightarrow^* Q$ and $\sim Q \rightarrow^* \sim P$. Contraposition is, however, merely an especially simple case of a causal loop. My approach, therefore, will be to establish the much more general result that causal loops involving the ancestral of the relation of direct causal necessitation are impossible, from which it will follow, as an immediate corollary, that contraposition is impossible.

What is a causal loop, in the present context? Situations of the following sort would certainly count as causal loops:

$$P \rightarrow^* P_1 \ \& \ P_1 \rightarrow^* P_2 \ \& \ldots \& \ P_n \rightarrow^* P.$$

But given that I am classifying contraposition as a case of a causal loop, causal chains of the sort just described cannot exhaust the class of causal loops. Rather, a causal loop is any situation of the following sort:

$$P \to^* Q_1 \ \& \ P_1 \to^* Q_2 \ \& \ \ldots \ \& \ P_{n-1} \to^* Q_n \ \& \ P_n \to^* P^*$$

where, first, P^* is identical either with P or with $\sim P$, and secondly, for every i, P_i is identical either with Q_i or with $\sim Q_i$.

The proof of this theorem is rather lengthy, but it may be helpful to sketch the basic method. Thus, if there were a causal loop

$$P \to^* Q_1 \ \& \ P_1 \to^* Q_2 \ \& \ \ldots \ \& \ P_{n-1} \to^* Q_n \ \& \ P_n \to^* P^*$$

the following probability would have to be defined:

Prob(Px, $P \to^* Q_1 \ \& \ P_1 \to^* Q_2 \ \& \ \ldots \ \& \ P_{n-1} \to^* Q_n \ \& \ P_n \to^*$ $P^* \ \& \ S$).

The idea is then to show that this probability is equal to another one where the probability of Px is conditional upon a shorter conjunction. This can be done by establishing the following four lemmas:

(1) Prob(Px, $P \to^* Q \ \& \ S$) = Prob(Px, $\sim P \to^* \sim Q \ \& \ S$);
(2) Prob(Px, $P \to^* Q \ \& \sim Q \to^* R \ \& \ S$) = Prob($Px$, S);
(3) Prob(Px, $P \to^* Q \ \& \ Q \to^* R \ \& \ S$) =
   Prob(Px, $P \to^* R \ \& \ S$);
(4) Prob(Px, $\sim P \to^* \sim Q \ \& \sim Q \to^* R \ \& \ S$) =
   Prob(Px, $\sim P \to^* R \ \& \ S$).

Lemmas (3) and (4) can be used to shorten the conjunction of causal statements upon which the probability of Px is conditional, while lemma (1) enables one to transform the initial conjunct when that is necessary before lemmas (3) and (4) can be applied.

What happens is that the reduction is carried out until the conjunction upon which the probability of Px is conditional consists of two causal statements, along with S. At that point one can show, using postulates (R_1^*), (R_2^*), and (R_3^*), along with lemma (2), that there are two different ways of working out the value of the resulting probability, and that they lead to contradictory results. Accordingly, causal loops are impossible.

The reader who is interested in the details of this argument will find a full statement of the proof in the Appendix.

Theorem 20: *Contraposition is Impossible*. The fact that it is logically impossible for it to be the case that both $P \rightarrow^* Q$ and $\sim Q \rightarrow^* \sim P$ is an immediate corollary of Theorem 19.

This completes the discussion of the formal development of the properties of the ancestral of direct causal necessitation. We have seen, first, that it must satisfy postulates having the same form as (R_1), (R_2), (R_3), and (R_6); second, that in addition to being transitive, it is also both irreflexive and asymmetric; third, that it is connected with increase in probability; and, finally, that not only contraposition, but causal loops in general, involving the ancestral, are logically impossible.

8.4.3 *Direct Causal Necessitation: Two Further Properties*

It was shown, in section 8.4.1, that the relation of direct causal necessitation is asymmetric and irreflexive, and that it is connected, in a certain way, with increase in probability. There are, however, two important properties of direct causal necessitation that were not dealt with at that point. These are, first, that contraposition necessarily fails, and second, that causal loops involving the relation of direct causal necessitation are impossible. These results follow immediately, however, from the discussion of the properties of the ancestral in the previous section.

Theorem 21: *Causal Loops are Impossible.* Any causal loop involving the relation of direct causal necessitation would entail the existence of a causal loop involving the ancestral of that relation. It therefore follows from Theorem 19 that there cannot be causal loops involving the relation of direct causal necessitation.

Theorem 22: *Contraposition Necessarily Fails.* It is an immediate corollary of the previous theorem that it is impossible, for any properties P and Q, for it to be the case both that $P \rightarrow_k Q$ and $\sim Q \rightarrow_k \sim P$.

8.4.4 *Causal Laws, Necessary and Sufficient Conditions, and Causal Priority*

The notions of a causal law, of causally sufficient and causally necessary conditions, and of causal priority, can now be defined as follows. First, the concept of a causal law:

It is a causal law that the possession of property P brings about the possession of property Q

means the same as

$P \to^* Q$.

Second, there is the concept of a causally sufficient condition, which can be defined in the same way:

Possession of property P is a causally sufficient condition of the possession of property Q

means the same as

$P \to^* Q$.

Third, there is the concept of a causally necessary condition. Since to say that the possession of property P is a causally necessary condition of the possession of property Q is just to say that the non-possession of property P is a causally sufficient condition of the non-possession of property Q, the concept of a causally necessary condition can be analysed as follows:

Possession of property P is a causally necessary condition of the possession of property Q

means the same as

$\sim P \to^* \sim Q$

Finally, there is the relation of causal priority. If the possession of property P is a causally sufficient condition of the possession of property Q, then the former is causally prior to the latter. The same is true if the possession of property P is a causally necessary condition of the possession of property Q. But the relation of causal priority may obtain in cases where the possession of one property is neither causally sufficient nor causally necessary for the possession of some other property. It might be the case, for example, that the possession of property P is a causally sufficient condition for the possession of property Q, and that the latter, in turn, is a causally necessary condition for the possession of property R. Then the possession of property P would be causally prior to the possession of property R, even though it need be neither causally sufficient nor causally necessary for the possession of R. The appropriate account of the relation (or, more accurately, quasi-relation) of causal priority would therefore seem to be as follows:

> Possession of property *P* is causally prior to possession of property *Q*
>
> means the same as
>
> Properties *P* and *Q* stand in the ancestral of the quasi-relation that obtains between two properties when possession of the one is either causally sufficient or causally necessary for the possession of the other.

Given these accounts, together with the properties of the ancestral of direct causal necessitation, a number of results immediately follow. First, by virtue of Theorems 17 and 18, and the fact that the ancestral is transitive, one has:

Theorem 23: Causal Laws are Transitive, Asymmetric, and Irreflexive.

Second, in view of Theorem 20, we have:

Theorem 24: For any properties P and Q, it cannot be the case both that possession of P is causally sufficient for the possession of Q, and that Q is causally necessary for the possession of P.

Finally, there is the following theorem concerning causal priority:

Theorem 25: Causal Priority is Transitive, Irreflexive, and Asymmetric. This can be established as follows. First, for the possession of property *P* to be causally prior to the possession of property *P*, one of the following would have to be the case:

(1) Property *P* stands in the relation of direct causal necessitation to itself;

(2) The 'negative universal', ~*P*, stands in the relation of direct causal necessitation to itself;

(3) There is a causal loop, involving the relation of direct causal necessitation, and one of whose elements is either *P* or ~*P*.

The first two possibilities are ruled out by the fact that direct causal necessitation is irreflexive, while the third is precluded by Theorem 21, concerning the impossibility of causal loops. So causal priority must be irreflexive.

For it to be the case both that possession of property *P* is causally prior to possession of property Q, and conversely, there would have to be a causal loop involving the relation of direct causal necessitation. So causal priority must be antisymmetric, and this, together with the fact that it is irreflexive, entails that it must be asymmetric as well.

Finally, given that causal priority is an ancestral, it is also necessarily transitive.

8.5 FORMAL DEVELOPMENT: II. THE INTERMEDIATE VIEW

Given the results in section 8.4, together with the relation between the intermediate and supervenience accounts, the present topic can be dealt with very quickly. For given that the two accounts differ structurally only with respect to one postulate—namely, the sixth—and that the intermediate version of that postulate is formally stronger than the supervenience version, theorems that are structurally identical with those established in section 8.4 must hold in the case of the intermediate approach.

It therefore follows from the intermediate account that both direct causal necessitation and its ancestral are irreflexive and asymmetric, that they are related to increase in probability, that contraposition necessarily fails, and that all causal loops involving these relations are impossible. The results for causal laws, causally necessary and causally sufficient conditions, and causal priority, carry over in comparable fashion.

There will also be some additional results not common to the two approaches, due to the fact that postulate (S_6) is stronger than postulate (R_6). Three of these are perhaps worth mentioning. The first is that the ancestral of direct causal necessitation satisfies an appropriate analogue of (S_6), namely:

(S_6^*): If $P \to^* Q$, then it is a law that, for all x, if Px, then Qx and C^*xPxQ

where the relation C^* is the ancestral of relation C.

The second result is similar to postulate (S_3), except that it concerns the probability, not that x has property Q, but that x has property Q as a result of x's having property P:

(S_7): Prob $(CxPxQ, P \rightarrow Q \& S) = \text{Prob}(Px, S)$,
      provided that S satisfies requirement T.

The proof of this is very short. First, by virtue of probability theory, we have that:

$\text{Prob}(CxPxQ, P \rightarrow Q \& S) =$
$\text{Prob}(CxPxQ, Px \& P \rightarrow Q \& S) \times \text{Prob}(Px, P \rightarrow Q \& S) +$
$\text{Prob}(CxPxQ, {\sim}Px \& P \rightarrow Q \& S) \times \text{Prob}({\sim}Px, P \rightarrow Q \& S).$

But, given that $CxPxQ$ entails Px, $\text{Prob}(CxPxQ, {\sim}Px \& P \rightarrow Q \& S)$ must be equal to zero. Moreover, by postulate (S_1), $\text{Prob}(Px, P \rightarrow Q \& S)$ is equal to $\text{Prob}(Px, S)$, while, by postulate (S_6), we have that $\text{Prob}(CxPxQ, Px \& P \rightarrow Q \& S)$ must be equal to one. Substituting these in the above equation then gives us the result that $\text{Prob}(CxPxQ, P \rightarrow Q \& S) = \text{Prob}(Px, S)$.

Finally, there is the analogue of (S_7) for the ancestral of direct causal necessitation:

$(S_7{}^*)$: Prob $(C^*xPxQ, P \rightarrow^* Q \& S) = \text{Prob}(Px, S)$,
      provided that S satisfies requirement T.

The proof of $(S_7{}^*)$ parallels the proof of (S_7), with appeal to (S_6) replaced by the use of $(S_6{}^*)$.

— 9 —

The Nature of Causation:
Further Issues

The first section of this chapter is concerned with the extension of the above approach to the probabilistic case. I shall attempt to show that this can be done in a straightforward fashion.

The second section is concerned with epistemological issues. There I shall attempt to show, not only that it is in principle possible to have evidence for causal judgements, but that the sorts of non-causal facts that we normally take to be evidentially relevant to such judgements are relevant, given the account of causation offered here.

The third section touches upon the difficult question of the relation between time and causation, and especially upon the prospects for a causal theory of time. A defence of such a theory is not possible here, but I do try to survey the main issues that would need to be dealt with in developing such an account, and to indicate the type of causal account that seems to me most plausible.

In the final section, the virtues of the present account of causation are briefly discussed.

9.1 CAUSAL RELATIONS AND PROBABILISTIC LAWS

9.1.1 *Issues and Alternatives*

As was indicated in section 5.7, the possibility of probabilistic laws appears to raise two important questions concerning causation. The first is whether causation can be combined with probabilistic laws at all. Is it possible for two states of affairs to be causally related when the underlying law is merely probabilistic?

As was also noted earlier, some philosophers, such as Hugh Mellor, maintain that the answer is no. The underlying train of thought, perhaps, is that the notion of a cause involves the notion of a complete cause, and that the latter, in turn, must be unpacked in terms of the concept of causally sufficient conditions. But where states of affairs are connected by merely probabilistic laws, there can be no causally sufficient conditions. Accordingly, the notion of causation cannot be combined with that of probabilistic laws.

There are a number of other, closely related grounds that might be offered for this view. One is that the notion of a causal relation between states of affairs involves the fundamental principle, 'Same cause, same effect'—a principle that is not compatible with there being causally related states of affairs that fall under merely probabilistic laws.

Alternatively, one can appeal to the notion that the connection between causally related states of affairs must have a force akin to that which is typically expressed by counterfactual conditionals. It is not enough to say that one state of affairs occurred, and was followed by another that stands in a certain relation to it. One needs to be able to say that given that the one state of affairs occurred, the other *could not* have failed to occur. But it is precisely this that one cannot assert if the underlying laws are merely probabilistic.

It would appear, however, that most present-day philosophers think that this view of the matter is mistaken, and that there can be causal relations in cases where the underlying laws are probabilistic. The most influential consideration in this regard has probably been quantum physics, since it seems to render plausible the idea that there could be a world containing causally related events, but where the laws are merely probabilistic. The most direct and forceful answer, however, to the claim that causation and probabilistic laws do not mix, would seem to be to set out an analysis that shows that states of affairs falling under probabilistic laws can be causally related. This is the tack to be taken here.

Assume, for the moment, that causation can be combined with probabilistic laws. The second issue then arises. Namely, are there probabilistic causal relations? That is to say, do causally related states of affairs falling under a probabilistic law necessarily stand in a different causal relation than causally related states of affairs falling under a non-probabilistic law?

The answer may depend upon the correct view of causal relations. If, contrary to what was argued above, a singularist view were right, there would be no distinct causal relations of a probabilistic sort. Causally related states of affairs would stand in the same causal relations regardless of whether they fell under deterministic laws, or probabilistic laws, or under no laws at all.

In contrast, the situation is less clear-cut, given the supervenience view, or the intermediate account. For, on the one hand, if causal relations are supervenient upon causal laws, together with non-causal states of affairs, or if they at least presuppose causal laws, one might take the fact that the relevant laws may be either probabilistic or non-probabilistic as grounds for holding that the causal relations in question are different. But, on the other hand, it might be argued that the existence of causally related events merely entails the existence of some law or other, and that since the form of the specific law does not matter, there is no reason to say that different causal relations are involved in the probabilistic and non-probabilistic cases.

9.1.2 *Probabilistic Causal Relations and the Supervenience View*

According to the supervenience view, two states of affairs stand in a probabilistic causal relation if and only if their non-causal properties, and the non-causal relations between them, are such that they fall under some probabilistic causal law. What is needed, therefore, is an account of probabilistic causal laws.

On the view advanced above, laws are second-order states of affairs involving genuine relations among universals. Let us introduce, then, the following notation:

$$P \to (k) \, Q.$$

The intended interpretation of this is that properties P and Q stand in what may be called the relation of direct causal probabilification to degree k, this being the relation between universals by virtue of which it is a basic causal law that if something has property P, then the probability that it also has property Q is equal to k.

It now needs to be shown that this expression can be defined. Given the discussion, in section 8.3, of non-probabilistic causal laws, the natural approach to this problem is to see whether there is some straightforward modification of postulates (R_1) to (R_6) that will produce postulates which will be satisfied by those second-order relations between universals that enter into basic probabilistic causal laws.

Postulates (R_1) and (R_2) state that the probability that something has property P is, in a specified range of cases, unchanged by the additional information that possession of property P results, either directly or indirectly, in the possession of property Q. The desired probabilistic analogues will simply state that the same is true with respect to the information that possession of property P has a certain *likelihood* of resulting in the possession of property Q. So we have the following postulates:

(N_1): $\text{Prob}(Px, P \to (k) \, Q \, \& \, S) = \text{Prob}(Px, S)$,
provided that S satisfies requirement T;

(N_2): $\text{Prob}(Px, P \to (k) \, P_1 \, \& \, P_1 \to (k) \, P_2 \, \& \, \ldots \, \& \, P_n \to (k) \, Q \, \& \, S) = \text{Prob}(Px, S)$,
provided that S satisfies requirement T.

Postulate (R_3), in contrast, specifies how the probability that something will have property Q is changed, in a certain range of cases, by the additional information that possession of some other property P results in the possession of property Q. What will be the probabilistic analogue of (R_3)? One very natural way of arriving at an answer is suggested by the fact that (R_3) is entailed by (R_1) together with (R_6). One can decide upon the probabilistic analogue of (R_6), and then use it, together with (N_1), to derive the probabilistic analogue of (R_3).

The most natural suggestion as to the probabilistic analogue of (R_6) would seem to be this:

(N_6): If $P \to (k) \, Q$, then it is a law that, for all x, the probability that Qx, given Px, is equal to k.

Given this, together with (N_1), the value of $\text{Prob}(Qx, P \to (k) \, Q \, \& \, S)$ can now be worked out as follows. First, it is a theorem that

$\text{Prob}(Qx, P \to (k) \, Q \, \& \, S) =$
$\text{Prob}(Qx, Px \, \& \, P \to (k) \, Q \, \& \, S) \times \text{Prob}(Px, P \to (k) \, Q \, \& \, S) +$
$\text{Prob}(Qx, {\sim}Px \, \& \, P \to (k) \, Q \, \& \, S) \times \text{Prob}({\sim}Px, P \to (k) \, Q \, \& \, S)$.

In view of (N_6), and a relevant property of probabilistic laws, $\text{Prob}(Qx, Px \, \& \, P \to (k) \, Q \, \& \, S)$ will be equal to k, while, by virtue of (N_1), $\text{Prob}(Px, P \to (k) \, Q \, \& \, S)$ will be equal to $\text{Prob}(Px, S)$, and Prob($\sim Px$, $P \to (k)$ Q & S) will be equal to Prob($\sim Px$, S). When these three substitutions are made, and one makes explicit the condition upon which they depend, namely, that S satisfies requirement T, we have the following candidate for the role of the probabilistic analogue of (R_3):

(N_3): Prob(Qx, $P \to (k)$ Q & S) =
$k \times$ Prob(Px, S) + Prob($\sim Px$, S) $\times$
Prob(Qx, $\sim Px$ & $P \to (k)$ Q & S),
provided that S satisfies requirement T.

Postulate (R_4) states that, given that an object lacks property P, the probability of its having some other property, R, is unaffected by the fact that it is a causal law that the possession of property P brings about the possession of property Q. It certainly seems plausible that the situation does not differ if the causal law in question happens, instead, to be probabilistic. So it would seem that the probabilistic analogue of (R_4) is acceptable:

(N_4): Prob(Rx, $\sim Px$ & $P \to (k)$ Q & S) =
Prob(Rx, $\sim Px$ & S),
provided that S satisfies requirement T.

Finally, just as (R_5) follows from (R_3) and (R_4), so the probabilistic analogue of (R_5) is derivable from (N_3) and (N_4):

(N_5): Prob(Qx, $P \to (k)$ Q & S) =
$k \times$ Prob(Px, S) + Prob($\sim Px$, S) $\times$
Prob(Qx, $\sim Px$ & S),
provided that S satisfies requirement T.

Given postulates (N_1) to (N_6), direct causal probabilification and the concept of a basic, probabilistic, causal law can be explained by formulating the relevant analytical theory of direct causal probabilification, and then by applying the familiar Ramsey/Lewis approach to theoretical terms. First, then, one has the following theory of direct causal probabilification:

The relation of direct causal probabilification to degree k, $\to (k)$, satisfies the following conditions:

(1) $\to$ (k) is a genuine, second-order relation between universals;

(2) for any particular x, any properties P, P_1, P_2, ... P_n, Q, and

R, and any statement S, → (k) satisfies the following postulates:

(N_1): Prob$(Px, P \rightarrow (k)\ Q\ \&\ S) =$ Prob$(Px,\ S)$,
          provided that S satisfies requirement T;
(N_2): Prob$(Px, P \rightarrow (k)\ P_1\ \&\ P_1 \rightarrow (k)\ P_2\ \&\ \ldots\ \&\ P_n \rightarrow (k)$
          $Q\ \&\ S) =$
          Prob$(Px,\ S)$,
          provided that S satisfies requirement T;
(N_3): Prob$(Qx, P \rightarrow (k)\ Q\ \&\ S) =$
          $k \times$ Prob$(Px,\ S) +$ Prob$(\sim Px,\ S) \times$
          Prob$(Qx, \sim Px\ \&\ P \rightarrow (k)\ Q\ \&\ S)$,
          provided that S satisfies requirement T;
(N_4): Prob$(Rx, \sim Px\ \&\ P \rightarrow (k)\ Q\ \&\ S) =$
          Prob$(Rx, \sim Px\ \&\ S)$,
          provided that S satisfies requirement T;
(N_5): Prob$(Qx, P \rightarrow (k)\ Q\ \&\ S) =$
          $k \times$ Prob$(Px,\ S) +$ Prob$(\sim Px,\ S) \times$
          Prob$(Qx, \sim Px\ \&\ S)$,
          provided that S satisfies requirement T;
(N_6): If $P \rightarrow (k)\ Q$, then it is a law that, for all x, the
          probability that Qx, given Px, is equal to k.

Given this theory of the relation of direct causal probabilification to degree k, that relation can then be explicitly defined by employing the Ramsey/Lewis approach. Direct causal probabilification to degree k will be defined as that relation which satisfies the relevant open formulas corresponding to conditions (1) and (2). Or, more accurately, it will be defined as the relation (or quasi-relation) that is the union of all of the relations satisfying the relevant open formulas—since it is possible, once again, that more than one relation satisfies the formulas in question.

Finally, one can say that it is a basic, probabilistic, causal law that the probability that something has property Q, given that it has property P, is equal to k, if and only if properties P and Q stand in the relation of direct causal probabilification to degree k.

9.1.3 *Probabilistic Causal Relations and the Intermediate View*

In the case of non-probabilistic causal laws and relations, we saw that, although the intermediate account of causation involves both a second-order relation between universals, and a first-order relation between states of affairs, whereas the supervenience account involves only the former, the two accounts are, in fact, extremely close, since they differ only with respect to a single postulate. We have also seen that, in the case of the supervenience approach, the account of direct causal necessitation can be transformed into an account of direct causal probabilification to degree k by means of a few simple and natural modifications. Given these two facts, it seems very plausible that parallel modifications in the intermediate account of direct causal necessitation will produce a satisfactory account of direct causal probabilification to degree k.

If we adopt this approach, we are led to the following, intermediate account:

The relation of direct causal probabilification to degree k, $\to (k)$, satisfies the following conditions:

(1) $\to (k)$ is a genuine, second-order relation between universals;

(2) for any particular x, any properties $P, P_1, P_2, \ldots P_n$, Q, and R, and any statements S, $\to (k)$ satisfies the following postulates:

(T_1): $\mathrm{Prob}(Px, P \to (k)\ Q\ \&\ S) = \mathrm{Prob}(Px, S)$,
provided that S satisfies requirement T;

(T_2): $\mathrm{Prob}(Px, P \to (k)\ P_1\ \&\ P_1 \to (k)\ P_2\ \&\ \ldots\ \&\ P_n \to (k)$
$Q\ \&\ S) =$
$\mathrm{Prob}(Px, S)$,
provided that S satisfies requirement T;

(T_3): $\mathrm{Prob}(Qx, P \to (k)\ Q\ \&\ S) =$
$k \times \mathrm{Prob}(Px, S) + \mathrm{Prob}(\sim Px, S) \times$
$\mathrm{Prob}(Qx, \sim Px\ \&\ P \to (k)\ Q\ \&\ S)$,
provided that S satisfies requirement T;

(T_4): $\mathrm{Prob}(Rx, \sim Px\ \&\ P \to (k)\ Q\ \&\ S) =$
$\mathrm{Prob}(Rx, \sim Px\ \&\ S)$,
provided that S satisfies requirement T;

(T_5): $\mathrm{Prob}(Qx, P \to (k)\ Q\ \&\ S) =$
$k \times \mathrm{Prob}(Px, S) + \mathrm{Prob}(\sim Px, S) \times$
$\mathrm{Prob}(Qx, \sim Px\ \&\ S)$,
provided that S satisfies requirement T;

(T_6): If $P \rightarrow (k)\ Q$, then it is a law that, for all x, the probability that $CxPxQ$, given Px, is equal to k.

The result here, as in the case of direct causal necessitation, is an account that differs structurally from the supervenience account only with regard to the final postulate, where, once again, the intermediate version of that postulate is formally slightly stronger than the supervenience version.

The overall conclusion, therefore, is that if, as I have urged, causation is fundamentally a matter of relations between universals which determine the direction of the logical transmission of probabilities, then there appears to be no problem in the idea of probabilistic causal relations. For, regardless of whether one accepts a supervenience view of causal relations, or an intermediate one, the characterization of probabilistic causal relations requires only very minor modifications in the accounts offered in the case of non-probabilistic causal relations.

9.2 EPISTEMOLOGICAL ISSUES

The fundamental idea, informally put, is that causation involves theoretical relations that determine the logical transmission of probabilities. What exactly this comes to depends, as we have seen, upon whether one adopts a supervenience view, or an intermediate view, of causal relations. But I have tried to show, in both cases, that it is possible to capture the intuitive idea, and to express it in a precise and simple way, and also that, when this is done, it can be established that causal relations, thus understood, possess the crucial formal properties that they are normally taken as having.

The situation, then, stands as follows. In the first place, the above accounts of causation appear to be formally adequate. In the second place, they also appear to be semantically adequate—that is to say, they appear to provide intelligible truth conditions for causal statements. The ground of this latter claim involves two contentions. One is that the approach to the meaning of theoretical terms suggested by Ramsey, and developed by Lewis, is a satisfactory one. The other is that this general approach can be applied to causal terms when causal relations are taken to be theoretical relations of the sort specified above.

If this is right, then one major issue remains: that of the epistemological adequacy of the above accounts. First, if causal relations are identified with the theoretical relations specified above, is it in principle possible to have knowledge, or at least justified beliefs, both about causal connections between particular states of affairs, and about causal laws? Second, if it is, what counts as evidence? Do non-causal facts of the sort that one normally takes to be evidentially relevant to causal claims turn out to be relevant, given the above accounts?

9.2.1 *Causal Claims: Direct Evidence and Indirect Evidence*

How can causal claims be justified? One possibility is that at least some causal beliefs might be non-inferentially justified. This view has, for example, been defended by Armstrong, who has argued, first, that one may very well have non-inferential knowledge that something is pressing on one's body, and second, that such knowledge is causal knowledge, since the concept of pressure involves causal notions.[1]

Is this possibility precluded by the view advanced above—that is, the claim that causal relations are theoretical? The answer will depend upon the correct account of non-inferential knowledge. Some accounts imply that it is impossible to have non-inferential knowledge of theoretical states of affairs. Others do not.

The epistemological issue is a difficult one. Fortunately, I think that it can be sidestepped. For even if it is possible to have non-inferentially justified beliefs concerning theoretical states of affairs, there would seem, as a matter of fact, to be relatively few beliefs concerning causal relations that present themselves as plausible candidates for membership in the class of non-inferentially justified beliefs. The vast majority of causal beliefs, surely, can only be justified on the basis of evidence.

Let us consider, then, what could count as evidence for causal claims. In approaching this question, I think that it will be helpful to draw a distinction between direct evidence and indirect evidence. Sometimes the evidence for a causal claim does not involve any further causal claims. Such evidence may be referred

---

[1] Armstrong, *A Materialist Theory of the Mind*, New York, 1968, p. 97.

to as direct evidence. In other cases, the evidence that two states of affairs are causally related involves information concerning other causal relations. Evidence of this sort may be referred to as indirect evidence.

Whether a certain type of evidence is to be classified as direct, or indirect, is not always an easy question to answer, since it is not always clear whether certain concepts involve causal components. The status of temporal information is especially problematic. But as I am inclined to believe that a causal account of temporal order and direction is correct, I shall here assume that direct evidence for causal claims cannot involve any information concerning temporal relations.

This distinction between direct and indirect evidence may be illustrated by the following. Suppose that one has information concerning the spatial arrangement of certain objects at different times, but no information at all concerning the temporal relations between those different, momentary states of affairs. In some cases it will be possible to arrive at highly probable conclusions concerning not only the causal ordering of those states of affairs, but also the direction of that order. In other cases, it will not be possible to arrive at such a conclusion. As an illustration of the former, consider the case of a stone striking a pond, resulting in outward-moving, concentric waves. Given information about the state of the surface of the pond at different times, together with knowledge of the relevant laws, it would be a straightforward matter both to order those instantaneous states, and to determine the probable direction of the causal processes involved.

Consider, in contrast, the case of a small number of particles moving about in an isolated region of space. Given information about the positions of the particles at different times, but none concerning the temporal relations between those instantaneous states of affairs, knowledge of the relevant laws may very well enable one to determine the structure of the causal network involved, but it will not enable one to determine the direction: each of the alternative directions will be about equally likely. If one is to be able, in such a case, to determine the direction of the causal processes, it would seem that additional information is required. In particular, it would seem that one must be able to relate the events in question to other events involved in causal processes whose directions are known.

9.2.2 *Direct Evidence: Apparent Control and Irreversible Processes*

There seem to be at least two especially important sorts of direct evidence. First, there is that provided by cases of apparent control. Second, there is the evidence provided by irreversible processes, both entropic and non-entropic.

Consider a simple case of apparent control. A pendulum to which a mirror is attached is swinging back and forth through the path of a light beam from a fixed source. Even if one had had no previous acquaintance with mirrors, one could quickly conclude that it was the presence of the mirror at a certain point which was causing the light to be reflected. What would be the basis of this judgement? The fact that the light alters direction when and only when the mirror is in a certain position is certainly relevant. But, by itself, this is not sufficient, since that fact is also compatible with the hypothesis that it is the light's changing direction at a certain point that causes the mirror to be there. There are, however, two other facts that are obviously relevant, and that point towards the hypothesis that it is the presence of the mirror that causes the light to change direction, rather than vice versa. The first is the fact that the frequency of the pendulum plus attached mirror is the same regardless of whether the light source is present or not. The second is the fact that if the frequency of the pendulum is altered, the frequency with which the light beam changes direction will also change in the same way. These facts appear to provide strong support for the idea that the frequency of the pendulum is determined by factors that are completely independent of the light source, and that it is the frequency of the pendulum that determines, in turn, the frequency with which the light beam changes direction.

What must now be shown is how such facts, which seem intuitively relevant to the causal judgement in question, turn out to be relevant given the account of causation set out above. To do this, it will suffice to consider the account of the relation of direct causal necessitation offered in section 8.3.2. Let the specification, S, of the background situation contain, first, statements of those laws that together with information about the nature of the pendulum and its location, determine its frequency, and second, a description of the experimental situation, including the location of the light source, the fact that the pendulum is free to swing back and forth, and the fact that it has a certain total energy, potential plus kinetic. Let us take P to be the property that a point has, at a given time, if part of the pendulum's mirror occupies that point at that time, and Q to be the property that a point has, at a given time, if a light beam changes direction at that point. Let a be the point at which the path of the pendulum intersects that of the beam of light. Finally, let n be the probability, given S, that a has property P, and let k be the probability, given S, that a has property Q. n will be a function of the period of the pendulum, the size of the mirror, etc., and will be substantially different from zero. k, in contrast, is presumably very small. Now consider the hypothesis that a light beam's coming into contact with a mirror causally necessitates a change in the direction in which the light is travelling. One consequence of this hypothesis, given the analysis of the relation of direct causal necessitation, can be seen by an application of postulate (R_5), which gives us:

$$\text{Prob}(Qa, P \rightarrow_i Q \ \& \ S) =$$
$$\text{Prob}(Pa, S) + \text{Prob}(\sim Pa, S) \times \text{Prob}(Qa, \sim Pa \ \& \ S).$$

If we assume that $\text{Prob}(Qa, \sim Pa \ \& \ S)$ is approximately equal to $\text{Prob}(Qa, S)$, we then have that $\text{Prob}(Qa, P \rightarrow_i Q \ \& \ S)$ is approximately equal to

$$n + k(1 - n).$$

If k is very small in relation to n, this will be very close to n. Thus, if the causal hypothesis is true, one would expect that the frequency with which light changes direction at point a will be very close to n. The value of n is determined, however, by information concerning the nature of the pendulum, the size of the mirror, etc., together with information about the relevant laws. It is possible to determine, then, whether light is changing direction at point a with the frequency to be expected given the causal hypothesis in question. If it is, that provides grounds for accepting that hypothesis. Thus, the presence of apparent control, given the analyses of causal notions offered here, does provide evidence for causal judgements.

Next, what about irreversible processes? Does it follow from the above analyses that such occurrences can provide evidence for causal claims? Consider the familiar example of apparent dispersal of order: concentric waves upon the surface of a pond. Let the specification, S, of the general situation include statements of the laws of motion and gravitation—but not of any other laws concerning the interaction of objects—together with information of a general sort concerning the existence of ponds and of objects moving about in their vicinity. Let us say that a region on the surface of a pond has property P at a given time if an object lands in that region at that time, and that it has property Q if some part of it is, throughout some short succeeding time interval, at the centre of a system of concentric waves. Let n be the probability that a region of a given size has property P, given that S, and k be the probability that it has property Q, given that S. n may not be especially large, but it seems clear that it will be much larger than k. For while it is true that if one were given detailed information about the nature of bodies of water, one could show that it was likely that concentric waves would arise when an object struck the surface, this will not be so given only the less detailed information contained in S. So given only S, any circular waves that arise will have to be viewed as doing so 'by chance', and the likelihood of this happening is extremely small. But now consider the hypothesis that an object's striking the surface of a pond causally necessitates the existence of outgoing circular waves throughout a succeeding time interval. Postulate (R_1) implies that $\text{Prob}(Px, P \rightarrow_i Q \,\&\, S)$ will be equal to n, while postulate (R_5) implies that $\text{Prob}(Qx, P \rightarrow_i Q \,\&\, S)$ will be very close to n, provided that $\text{Prob}(Qx, {\sim}Px \,\&\, S)$ is approximately equal to $\text{Prob}(Qx, S)$. So if the observed frequencies do have roughly those values, that will provide evidence for the causal hypothesis in question. Apparent dispersal of order is, on the account of causality offered above, relevant evidence.

Another important class of phenomena that exhibit irreversibility in an obvious way consists of events in which something leaves a *trace*. Consider, for example, the sequence of events that takes place when a hot stamp comes into contact with a piece of wax, and then moves away, leaving an impression in the wax of the same size and shape as the stamp. If this sequence of events were viewed as a causal process running in the reverse direction, extreme improbability would be involved: that the stamp and the impression should be of exactly the same size and shape; that the stamp should be moving in just such a way that it fits perfectly into the impression; that on moving away it leaves the wax both smooth and cool, and so on. In contrast, when viewed as a causal process in the normal direction, there is nothing especially improbable about such a sequence of events, since the perfect correspondence of stamp and impression will, by virtue of postulate (R_5), have a probability of the same order as that of a warm stamp's coming into contact with a piece of wax. The present account of causation explains, therefore, why apparent traces provide evidence for causal judgements.

What is true with respect to processes that involve apparent traces, or apparent dispersal of order, holds for irreversible processes in general. For the reason that irreversible processes virtually always run in one direction is that they involve a causal connection between two sorts of events, one of which is, antecedently, reasonably probable, and the other of which is, antecedently, extremely improbable. If there were a sequence in which the antecedently extremely improbable event was causally prior, that event would remain just a grand accident. In contrast, if the antecedently very improbable event is causally posterior to the other event, then by virtue of postulate (R_5) it ceases to be a highly improbable event: its probability cannot be less than that of the other, more probable event. The view that causal relations are relations that determine the direction of transmission of probabilities therefore makes it possible to see why it is that irreversible processes provide evidence for causal claims.

9.2.3 *Indirect Evidence*

States of affairs that provide direct evidence for causal connections abound, as is illustrated by the propagation of spherical wave fronts, by the resistance that bodies encounter as they move through a medium—due to the fact that there are more collisions with particles on the side in the direction in which the object is moving than on the other side—and by many other types of irreversible processes, both entropic and non-entropic. But there are also many cases where we believe that certain events are causally connected, even though there is no direct evidence that this is so. How can such beliefs be justified?

A plausible answer is provded by a variation on the 'nomological net' idea of Reichenbach, discussed in section 7.4. First, given that there are many events for which there is direct evidence that they are causally connected, and given that these events fall under certain laws, one is justified in holding not only that certain laws exist, but that they are causal laws. Second, one can then use those laws to place events in a causal network. This will enable one to determine, either that two events are causally related, or that one event is causally between two other events. Thus one may be able to conclude that either event A is causally prior to event B or vice versa, even though there is no direct evidence for either hypothesis by itself. Or, one may be able to conclude that event B is causally between events A and C, without knowing whether A caused B, and B caused C, or, alternatively, C caused B, and B caused A. In the latter sort of case, one will know that two causal processes have the same direction, without knowing what that direction is.

Suppose, finally, that in a case where one has evidence that event B is causally between events A and C, it is also true that B is a part of a spherical wavefront that originated in the vicinity of A. Given this direct evidence that A is causally prior to B, one can also conclude that B must be causally prior to C. It can be shown, therefore, that two states of affairs stand in a certain causal relation, even if there is no direct evidence that this is so.

This proposal differs from Reichenbach's in at least two respects. First, it does not involve the dubious principle of the local comparability of the time order. Second, and much more important, it is an answer to an epistemological question, not an ontological one. On Reichenbach's account, what makes it the case that event A is causally prior to event B involves, in some cases, events that are either causally prior to A, or causally posterior to B. I have contended that this idea—that the truth-maker for the statement 'Event A is causally prior to event B' may involve events that are not part of the causal process leading from A to B—is strongly counter-intuitive. The present proposal, however, does not involve this idea. Whether A is causally related to B can never depend, given the present account of causation, upon events that are either causally prior to both A and B, or causally posterior to both A and B. The suggestion is, instead, epistemological: information concerning events that are causally prior to both A and B, or causally posterior to both, may provide evidence concerning the direction of the causal process connecting A and B.

9.3 CAUSATION AND TIME

One very fundamental philosophical issue connected with causation concerns its relation to time. Do causal concepts presuppose temporal ones? Or vice versa? Or are they mutually independent?

A causal analysis of temporal concepts provides, I believe, the right answer. Such an analysis presupposes that a satisfactory account of the nature of causation can be given which is free of all reference to temporal relations. I have tried to show, however, that this can be done. The door is therefore open to a causal theory of time.

There are, however, other obstacles that stand in the way of such a theory, some of them quite serious.[2] But I want to suggest that some of the crucial objections arise, in fact, because of the kind of causal theory typically advanced, and I shall be proposing, in particular, that the most promising form for a causal theory of time is one where, first, a tensed, rather than a tenseless view of time is adopted, and second, a relational view of space and time is rejected in favour of an absolute conception.

9.3.1 *Two Views of Time: Tensed and Tenseless*

Temporal concepts fall into two distinct classes—tenseless and tensed. Tenseless temporal concepts pick out relations that define a temporal ordering of events in the world. Thus, among the central tenseless temporal concepts are those of simultaneity and temporal priority.

Tensed temporal concepts are also associated with a temporal ordering of events, but what distinguishes them from tenseless temporal concepts is that the temporal ordering is generated by relating events to the present. Among the central tensed temporal concepts, therefore, are the concepts of past, present, and future.

Related to this distinction between two types of temporal concepts are two views of time. According to the tensed view, the concepts of past, present, and future are central to the idea of time, and these concepts reflect ontologically fundamental aspects of the world. Advocates of a tensed view do not, however, always

---

[2] For a clear exposition of a number of objections to causal theories of time, see Smart, 'Causal Theories of Time', *Basic Issues in the Philosophy of Time*, ed. Freeman and Sellars, La Salle, Ill., 1971, pp. 61–77.

agree as to precisely what those aspects are. For there are, in fact, two quite distinct alternatives within the general tensed position. Thus, some advocates of a tensed approach to time maintain that the significant feature consists in the fact that the past, the present, and the future are not all real. (Sometimes it is held that both the past and the present are real, but the future is not, and sometimes, more radically, that only the present is real.) Others claim, however, that the past, the present, and the future are all equally real, but that there are intrinsic properties, of pastness, present-ness, and futurity, precisely one of which is possessed by any given event at any particular time.

According to the tenseless view, on the other hand, the concepts of past, present, and future do not point to anything ontologically significant. These concepts function, instead, simply to describe events from the persepctive of a given observer. They can be compared, then, to terms such as 'here' and 'there'. Just as the latter merely indicate the spatial location of an object relative to the speaker, and do not function to attribute any intrinsic properties to things in the world, so terms such as 'past', 'present', and 'future' merely indicate an event's temporal position relative to the speaker. The reality involved in time is simply that of the temporal order of events. The basic temporal concepts, therefore, are not those of past, present, and future, but those of temporal priority and simultaneity.

One question that needs to be considered, then, in setting out a causal theory of time, is whether one should opt for a tensed account, or for a tenseless one. The vast majority of philosophers who favour a causal theory of time also accept, I believe, a tenseless view of time. Why is this so? The answer, perhaps, is that, on the one hand, causal analyses typically begin with the tenseless concepts of simultaneity and temporal priority, while, on the other hand, advocates of a tensed view of time almost invariably hold that tenseless temporal concepts can be analysed in terms of tensed ones. Thus it has been proposed, for example, that A's being earlier than B can be analysed as its being the case that either it was, or it is now, or it will be, the case that A is past while B is present.

I want to suggest, however, that such analyses cannot be sustained, on the ground that the concepts of the past, the present, and the future cannot all be taken as analytically basic. In support,

I would appeal to the idea that analytically basic concepts must apply to things by virtue of directly observable properties and relations. For, given this thesis, it seems clear that the concept of the future cannot be analytically basic, since the fact that an event lies in the future is not something that, as things stand, we can know by direct observation. It is also very doubtful whether the concept of the past can be taken as analytically basic, although that is something that would need to be considered much more carefully.

But if the concept of the future, at least, is not analytically basic, how is it to be analysed? If causal concepts are not brought in immediately, it would seem that there is no alternative to the view that the future consists, by definition, of whatever is later than the present. If so, the conclusion is that, rather than tensed concepts serving as the basis for an analysis of temporal priority, the latter relation is itself needed to provide an analysis of at least some of the tensed concepts.

In short, the contention is that advocates of a tensed view of time are mistaken in thinking that they have at hand resources for analysing the concept of temporal priority, beyond those available to defenders of a tenseless account of time. As a result, a causal theory of time should have no less appeal for one who believes that tenses are ontologically significant than for one who does not.

Which view of time is more likely to be correct—the tensed, or the tenseless? This issue cannot be resolved here, but there are some points that should be made. In the first place, then, the view that there is a significant ontological difference between the past and the future—the former being fixed, and the latter open— seems to be a very widespread and deeply held belief about the world. Commonsense is, of course, badly mistaken about many things, and this may be merely another illustration. Still, I think that one is probably justified in holding that reasonably weighty considerations are needed if the tensed view is to be set aside.

Serious arguments against the tensed view have certainly been advanced. Of these, perhaps the most impressive are, first, McTaggart's famous attempt to show that the tensed view of time involves a contradiction,[3] and secondly, an argument from the special theory of relativity, to the effect that, since simultaneity is

---

[3] McTaggart, *The Nature of Existence*, ii, Cambridge, 1927, bk. 5, chap. 33.

relative to a frame of reference, there cannot be, at a given time, any absolute division of events into those that are past, those that are present, and those that are future; and that, given that this is so, there cannot be any ontological gulf which corresponds to tensed concepts.

Both arguments are formidable, but I believe that they can be answered. In the case of the argument from special relativity, it may be that one must argue, first, that there can be a supplemented version of relativity theory which incorporates relations of absolute simultaneity, and which is empirically equivalent to the unsupplemented theory; and secondly, that there are reasons for preferring the less austere theory.

In the case of McTaggart's argument, the appropriate conclusion, I think, is that a coherent formulation of a tensed view of time cannot make do with the notion of truth *simpliciter*: the concept of truth *at a time* is crucial. It therefore needs to be shown that a satisfactory analysis of the latter concept is forthcoming.

Perhaps, then, such objections to the commonsense belief in a tensed view of time can be answered. But is there any reason, in the present context, to be concerned about which view of time is correct? Does the choice have any bearing, either upon the topic of causation, or upon that of laws?

There are, I think, at least two connections. First, we ordinarily think of causes as *bringing about* their effects. Perhaps this intuitive way of talking about causes is fully captured by the idea of a logical transmission of probabilities. However, I am inclined to think that something of significance is added if one can combine the above account of causation with both a causal theory of time, and an appropriate tensed account. For, given a causal theory of time A's being the cause of B entails that A is earlier than B, while given a tensed view of time according to which the future is not real, it will then follow that there must have been a time at which A was real, while B was not. Given this result, it seems very natural indeed to speak of causes as bringing about their effects. The suggestion, in short, is that real causation may require real time.

The other connection concerns the nature of laws, and, in particular, the choice between the account developed here, and that defended by David Armstrong. Consider a world, one of whose laws relates properties, some of which were nowhere

instantiated before a certain time. For concreteness, suppose that
it is a psychophysical law to the effect that, whenever an organism
is in a brain state of type B, that causes it to have an experience of
the green variety, and suppose, further, that t_0 is the first time that
any organism had such an experience. What is one to say about the
following counterfactual: 'If an organism had been in a brain state
of type B at some time prior to t_0, it would have had an experience
of the green variety'? For this counterfactual to be true, the
relevant law must obtain, and if laws are relations among
universals, the appropriate universals must exist. But then, if one
holds, as Armstrong does, that uninstantiated universals cannot
exist, it follows that one cannot accept a tensed view of time of the
sort that asserts that the future is not real. For this would imply,
for one of the relevant universals—namely greenness, as a quality
of experiences—that the tenseless sentence asserting its existence
would not be true at any time prior to t_0. Armstrong's account of
laws requires, therefore, either a tenseless view of time, or,
alternatively, a tensed view which accepts the reality of the future.
The account set out above, in contrast, would seem to be
compatible with any view of the nature of time, including tensed
views which maintain that, while the past and present are real, the
future is not.

9.3.2 *An Absolute or a Relational Account?*

Causal theories of time have almost always been relational in form.
Rather than treating space-time as a substance, as something
capable of independent existence, they have viewed talk about
space and time as being replaceable, without loss, by talk about
spatio-temporal relations among physical objects and events.

   The main reason, perhaps, is that most philosophers have been
reluctant to countenance the idea of causal relations between
spatio-temporal points or regions, and it is precisely this idea that
is needed if one is to set out a causal theory of time that is not
relational in form. For to accept an absolute view of space and
time is to hold that space and time might exist even if there were
no physical objects or events, whilst to accept a causal theory of
time is to hold that temporal relations are analysable in terms of
causal ones. Together, these views imply that spatio-temporal
regions can stand in causal relations to one another, since

otherwise the idea of an empty spatio-temporal world would be incoherent.

But is there any good reason for ruling out causal relations between spatio-temporal points or regions? I do not believe that there is, and I want to suggest that the widespread failure to consider the possibility of such causal relations reflects a tendency to focus upon a limited range of causal situations—namely, cases of causal *interaction*, of causal processes that involve change. But static situations can also involve causal relations. Many philosophers have suggested, for example, that the identity of objects over time—including unchanging objects, such as fundamental particles—is most plausibly analysed in terms of causal relations between an object's temporal parts.

When one does focus upon the causal relations involved in such unchanging states of affairs, there no longer appears to be anything especially problematic in the idea that the existence of a spatio-temporal region at one time might be, for example, a causally sufficient condition of the existence of corresponding regions at later times.

If this line of thought is tenable, then it is certainly possible to combine a causal theory of time with an absolute view of space and time (or space-time). But I want to advance a stronger claim— namely, that this is the preferred form for a causal theory of time.

One of the more important considerations in support of this contention emerges if one considers the question of whether it is possible to set out a causal theory of time for special relativity. The case for an affirmative answer rests mainly on the work of Alfred A. Robb, who, in his profound and difficult book, *A Theory of Time and Space*, attempted to establish the remarkable conclusion that all of the qualitative and quantitative properties of the space-time of special relativity can be defined in terms of the single, purely qualitative relation of causation.[4]

Robb's discussion may be viewed as falling into two parts. The first, which is very brief, involves offering a causal account of the relation of *being after*. The second, which takes up almost all of the book, consists in his demonstration that, starting out from the

---

[4] Robb, *A Theory of Time and Space*, Cambridge, 1914. A more accessible account of Robb's approach is contained in his later work, *The Absolute Relations of Time and Space*, Cambridge, 1921, in which he outlines his theory without setting out the formal proofs.

supposition that there is a set of elements, some of which are after other elements, and which satisfy certain postulates, it is possible to develop, in a rigorous fashion, all of the geometry of Minkowski space-time. The demonstration requires the proof of more than 200 theorems, based upon 21 postulates, and is over 350 pages in length. Not surprisingly, Robb is sometimes referred to as the Euclid of relativity theory.

Concerning the soundness of the second, and mathematical part of Robb's work, there can certainly be no question. In itself, however, that part provides no reason for thinking that a causal account of the temporal concepts of the special theory of relativity can be given. What it establishes is the admittedly remarkable, but conditional conclusion, that *if* a causal account can be given for the purely qualitative relation of being after, then an account can be given for all other temporal concepts, both qualitative and quantitative. One is left, therefore, with the crucial question of whether it is possible to offer a satisfactory causal analysis of the basic qualitative relation of being after.

The analysis which Robb proposes for the latter relation is as follows: 'If an instant *B* be distinct from an instant *A*, then *B* will be said to be after *A*, if, and only if, it be abstractly possible for a person, at the instant *A*, to produce an effect at the instant *B*.'[5] The problem with this analysis is that the presence of the notion of effects that it is *possible* to produce, means that the relation of being after is getting explained in modal terms, so that if ultimate modal facts are not to be countenanced, appropriate truth-makers must be specified for statements of the sort in queston. What could those truth-makers be? What would make it the case that it is possible at *A* to produce an effect at *B*? Clearly, it cannot be a matter simply of there being certain laws, nor of that together with the possession by *A* and *B* of relevant intrinsic properties. It must also be a matter of an appropriate relation between *A* and *B*—and in particular an appropriate spatio-temporal relation. In short, when one attempts to supply non-modal truth-makers for the relevant modal propositions, it turns out that one needs to refer to the spatio-temporal relation which obtains between the locations

---

[5] Robb, *A Theory of Time and Space*, p. 7. The term 'instant', as ordinarily used, presupposes a relation of absolute simultaneity. Robb uses the term, however, simply to mean a space-time point.

in question. Robb's proposed definition of the relation of being after is, therefore, implicitly circular.

One response to this difficulty would be to modify Robb's definition by replacing the concept of causal connectibility by that of actual causal connection:

> If a space-time point B is distinct from a space-time point A, then B will be said to be after A if, and only if, some event at A produces an effect at B.

But this modified analysis suffers from at least two defects. First, there can surely be spatio-temporal locations, A and B, such that B is after A, but where there are no events taking place at either location. The modified analysis, in contrast to Robb's account, has the unwelcome implication that this is impossible.

Secondly, the modified account is not even satisfactory in the case of locations where events do occur. For suppose that it is possible, at A, to produce an effect at B. Then it is surely true that B is after A regardless of whether there are any causal connections between events that actually occur at A and B.

It may seem that we have encountered an insurmountable obstacle, since it appears that neither causal connectibility, nor actual causal connectedness, will do. I believe that this is in fact the case, *if* what one wants is a causal theory of a relational sort. In contrast, however, if one is willing to embrace absolute space-time, a solution appears to be at hand. For then one can formulate a causal theory in terms of actual causal connections, not between events in space-time, but between space-time points themselves.

In addition to providing a solution to the present problem, this approach enables one to answer a number of other familiar objections to causal theories of time. It has been argued, for example, that causal theories of time cannot make sense of the possibility of space-time points at which there are no events, nor of the more dramatic possibility of *times* at which nothing exists. It has also been objected that causal theories of time cannot allow for the possibility of 'points of space-time which are occupied by events which are neither effects nor causes of other events'.[6] Finally, there is the objection that causal theories of time are incompatible with a possibility that is allowed by the general

---

[6] Smart, op. cit., p. 63. The other objections mentioned in this paragraph have also been advanced by Smart.

theory of relativity, namely, that of totally empty spatio-temporal worlds. These objections provide, I believe, good grounds for rejecting any relational version of a causal theory of time, but they pose no problem for the alternative approach recommended here.

To sum up briefly, I believe it is one of the merits of the present account of causation that it does not preclude a causal theory of time. Whether a satisfactory causal analysis of temporal concepts can be developed is, however, a very difficult question, and one that I have not attempted to answer. What I have tried to do is to point to some important options that have generally gone unnoticed. Thus I have argued that a causal theory of time need not be relational, and I have tried to indicate some of the advantages of a causal theory which involves a substantival view of space-time. I have also argued that, contrary to what many believe, a causal theory of time coheres just as well with a tensed conception of time as it does with a tenseless view. This is, I believe, a welcome conclusion, since it provides grounds for somewhat greater optimism concerning the possibility of a theory of time which will enable one to make sense of two widely shared, pre-theoretical beliefs—namely, the belief that there is a very great ontological chasm between past and future, and the belief that causation involves the bringing into existence of an effect.

9.4 THE CASE FOR THE PRESENT APPROACH

What reasons are there for accepting the account of causation developed above? There are, perhaps, four main considerations. In the first place, the present account enables one to establish that causal relations have the formal properties they are normally taken to have. The relation of direct causal necessitation, its ancestral, the relation of being a causally sufficient condition, and the relation of being a causally necessary condition, can all be shown to be necessarily asymmetric, and all but the first is necessarily transitive. Moreover, the relation (or quasi-relation) that obtains between two states of affairs when one is either a cause of, or a causally sufficient condition of, or a causally necessary condition of, the other, is also necessarily asymmetric, as is the ancestral of that relation. As a consequence, it follows from the present account that it is logically impossible for there to be causal loops of states of affairs that are interconnected by the relation of being a cause of, or by the relation of being a causally sufficient condition of, or by the relation of being a causally necessary condition of, or by any combination of these.

In the second place, it seems logically possible that there could be causal processes that involved discontinuity, either spatially, or temporally, or both. Some accounts of causation preclude such possibilities. The present account does not.

In the third place, some arguments were set out, in Chapter 7, that appear to tell against all reductionist accounts of causation, aside from those that base the direction of causation upon the direction of time. I have in mind here the arguments concerning, first, the possibility of inverted universes, and secondly, the possibility of very simple universes. The thrust of the former argument was that it is possible for there to be universes that are, so to speak, temporal mirror images of each other. If this is right, it means that it is possible for the direction of causation to be opposite to the directions defined by such things as increase in entropy, apparent dispersal of order, open forks, both conjunctive and interactive, and apparent control. More generally, the 'inverted universes' argument would seem to show that it is logically possible for there to be two worlds that differ causally, even though they have exactly the same laws, and are indistinguishable with respect to all non-nomological, non-causal, properties and relations—with the possible exception of temporal ones, if it should turn out that the direction of time cannot be analysed in causal terms.

The thrust of the 'simple universes' argument was that it is logically possible for there to be certain very simple universes which contain causally related events, but which are devoid of certain other characteristics which are appealed to in many reductionist accounts of causation. In particular, I argued that there could be very simple universes in which there would never be any change in entropy, nor any apparent dispersal of order, nor any open forks, either conjunctive or interactive, but which nevertheless contained causally related states of affairs.

Each of these arguments radically restricts the range of possible reductionist accounts of the direction of causation. The only alternatives not precluded are those that analyse the direction of causation in terms of the direction of time. It is possible, of course, that the intuitions to which these two arguments appeal are, in the final analysis, unsound, but in the absence of specific reasons for thinking that this is so, the view that inverted universes, and simple universes, of the sort described, are genuine logical possibilities, would appear to be a reasonable one. Accordingly, it would seem to be a positive feature of the present approach that it does allow for such possibilities.

Finally, the realist account of causation set out above appears to be epistemologically satisfactory. For, as was argued in section 9.2, it can be shown not only that it is logically possible to have epistemically justified beliefs concerning causal states of affairs, but also that the sorts of considerations that one normally takes as evidence for causal claims are epistemically relevant. In particular, it can be shown that the direction of increase in entropy, the direction of non-entropic, irreversible processes, the direction defined by open forks, especially those of high complexity, and the direction of apparent control, all provide good evidence concerning the direction of causation.

To sum up: the view that causation, rather than being reducible to non-causal facts, involves theoretical relations which determine the direction of the logical transmission of probabilities, appears to have the following virtues. First, it seems to generate the desired formal properties of causal relations. Second, it enables one to make sense of what appear to be genuine logical possibilities, concerning discontinuous causal processes, 'inverted universes', and 'simple universes'. Finally, it does this without making causal relations epistemologically inaccessible. The case for a realist approach to causation is, I believe, a strong one.

PART IV:

A Summing Up

— 10 —

Summary and Conclusions

The accounts offered here of the nature of laws, and of causation, are, I believe, fundamentally empiricist in orientation. At the same time, it is clear that both accounts diverge rather radically from the accounts that empiricists have usually advanced. The source of this divergence is that traditional empiricist accounts of laws and causation are reductionist, whereas the present accounts are not.

What are the grounds, then, for classifying the foregoing accounts as empiricist? Essentially, there are two considerations. First, one of the central empiricist contentions is that necessary truths are analytic, and non-empiricist accounts of laws, and of causation, generally involve a rejection of this claim—it being maintained either that laws of nature are themselves necessary, or that certain general principles, such as that every event has a cause, are necessary truths. According to the present approach, however, laws of nature are merely contingent. Nor is it necessary that events have causes. In this respect, then, the present accounts fall squarely within the empiricist camp.

Second, another central empiricist doctrine is that terms that do not refer to what is immediately given in experience must, if they are to be meaningful, be capable of being analysed in terms of vocabulary that does satisfy this condition. Are the accounts compatible with this view? The answer depends upon how this semantical requirement is to be interpreted. In particular, one needs to know whether it is to be construed in such a way as to admit quasi-logical, or topic neutral vocabulary, as semantically basic.

Consider, for example, terms such as 'property', 'event', and 'relation'. If one is an empiricist, can one hold that such terms are meaningful, even if it turns out to be impossible to offer an analysis of them? I would argue that one can, on the grounds that, since one does have immediate experience of properties, events, and relations, the above terms apply to entities that are immediately given in experience. Some empiricists, however, have rejected this line of thought, and have held that, in view of the fact that all the properties, events, and relations that one ever experiences are, by definition, such as can be experienced, one cannot form general concepts—of properties, events, and relations—that would include, in their extension, entities that are incapable of being experienced: one can form at most the rather more modest concepts of properties, events, and relations that are *capable of being experienced.*

There would seem, however, to be at least two reasons for rejecting this contention. In the first place, I believe that this line of thought, rigorously pursued, leads to a conclusion that few will find acceptable. For it would appear that it can equally well be argued that, since all the properties, events, and relations that I experience are, by definition, ones that are experienced by *me*, I cannot form even the general concepts of properties, events, and relations that are capable of being experienced *by someone or other*. Moreover, though the argument is slightly more complicated, I believe that one can also show that this line of thought leads, in the end, to a semantical solipsism of the present given moment, according to which one can only understand terms that refer to what one is *now* experiencing.

In the second place, there is the question of whether empiricism allows one to make sense of statements about theoretical entities, realistically conceived. It seems to me that, if the inclusion of topic neutral, or quasi-logical terms among one's semantically basic vocabulary were incompatible with empiricism, then empiricism could not provide an interpretation of theoretical terms which would allow for the possibility of realism with respect to theoretical entities. In contrast, and as we saw in Chapter 1, a realist view of theoretical entities can be shown to be intelligible, if quasi-logical vocabulary is admissible.

In short, I believe there are good grounds for holding that any acceptable version of empiricism must admit, as semantically basic, topic neutral vocabulary. But if this is right, then the present accounts of laws, and of causation, in no way conflict with the fundamental semantical thesis of empiricism. For, in the first place, I have argued in support of the traditional empiricist claim that causal and nomological concepts do stand in need of analysis; and, in the second place, I have attempted to show how those concepts can be analysed if one has at one's disposal quasi-logical vocabulary.

Empiricist accounts of laws and causation have almost invariably been reductionist in nature. That is to say, it has been held, in the case of laws, that once all non-nomological properties and relations are fixed, then it is logically settled what laws there are, and, in the case of causal relations, that once all non-causal properties and relations are fixed, then it is logically determined how events are causally related. This pervasive feature of empiricist accounts is not, I believe, an accident. It seems to me intimately tied to the semantical issue just discussed. For if one accepts the claim that terms not referring to what is immediately given in experience must be analysable in terms of vocabulary that does meet this condition, and then fails to consider the possibility that the semantically basic vocabulary may include quasi-logical terms, one will inevitably be driven to reductionist accounts of laws and of causation.

In contrast, if one does accept quasi-logical, or topic neutral vocabulary, the way is clear for realist accounts of both laws and causation. I have tried to show, moreover, that there are good reasons for holding that reductionist accounts, in both areas, must be abandoned in favour of realist ones. One of the basic lines of argument, in the case of laws, was that it seems possible to imagine worlds that differ with respect to what laws they contain, but which happen not to differ with respect to any non-nomological events or states of affairs. It would seem, for example, that one can imagine two worlds that contain different probabilistic laws, but that turn out to be indistinguishable as long as one considers only the non-nomological properties of, and relations among, the events that take place. It would seem, too, that there could be a world that was completely anomic, but which, by a grand accident, resembled exactly, with respect to the non-nomological properties of, and relations among, the events in that world, a world where events fell under deterministic laws. Similarly, in the case of causation, it would seem that one can imagine worlds that, although they are identical with respect to all the non-causal properties and relations of corresponding events, nevertheless differ with respect to the causal connections between events. Realist approaches to laws and causation enable one to make sense of such possibilities; reductionist accounts do not.

A realist account of laws appears to call for realism with respect to universals, since it is not easy to see what laws can be, on a realist view, other than relations among universals. But that step having been taken, one is confronted with some important options. One, which I have not discussed here, is the choice between accounts that maintain that the relations in question are necessary ones, and those that hold that they are contingent. I have simply followed the traditional empiricist view that laws of nature are contingent.

Another important choice is between accounts that rest upon an Aristotelian view of universals—according to which all universals must be instantiated—and accounts that involve a view of universals that is at least partially Platonic, in the sense of countenancing the possibility of at least some uninstantiated universals. The former sort of theory has been championed by Armstrong, while I have supported the latter alternative. The issues that are at stake in the choice between these alternatives were described at the beginning of Chapter 2, and discussed further in later sections. As we saw, the great merit of Armstrong's approach appears to be its compatibility with a thoroughgoing naturalism. Nevertheless, it seems to me that there is reason for preferring the more Platonic version of the theory that has been advanced here, on the grounds, first, that the latter coheres better with our intuitions concerning what sorts of laws there could be, and especially with our intuitions concerning the possibility of uninstantiated laws; second, that it makes possible the sort of account of the intrinsic nature of nomological relations which was advanced in section 3.2; and third, that it is, in contrast to Armstrong's account, compatible with all views of the nature of time, both tensed and tenseless.

Two other issues discussed in Part II deserve to be mentioned. The first concerned the possibility of justified beliefs about what laws of nature there are. At first glance, it might seem that it would be easier to justify such beliefs if one accpeted a reductionist account. But this is not in fact so. For, as a number of writers have noted, if laws are identified with cosmic regularities, then it is hard to see how, in a large universe, one could ever have evidence which made it at all likely that some exceptionless regularity obtained. In contrast, as I attempted to show in section 3.3, if laws are identified, not with regularities, but with relations among universals that underlie and explain those regularities, then one can have evidence which will strongly confirm hypotheses concerning what laws there are.

The other issue concerned the possibility of probabilistic laws. This is an important question since, given quantum physics, it would seem that one could hardly accept an account of laws which did not allow for the possibility of probabilistic laws. This issue was dealt with in Chapter 4, where I attempted to show, first, that reductionist accounts cannot provide adequate truth-makers for statements expressing probabilistic laws, and second, that the realist account of non-probabilistic laws set out in Chapters 2 and 3 can be extended, in a relatively straightforward manner, to the case of probabilistic laws.

The discussion of the nature of causation in Part III involved four main tasks. First, I tried to show, in Chapter 7, that the reductionist accounts of causation which philosophers have advanced are all exposed to serious objections. In doing so, however, I attempted to point to certain general considerations that tell against almost any reductionist account of causation. Those general considerations involved worlds that differed with respect to causal connections between events, yet which did not differ with respect to any of the non-causal properties of, or relations between, those events.

Second, I set out a detailed account of the nature of causation. The intuitive idea underlying that account was that causal relations between states of affairs are relations that, so to speak, determine the direction of the logical transmission of probabilities, and, thereby, the likelihoods of different types of events. I attempted to show how this intuitive idea can be given a precise formulation that involves a small number of postulates which specify how the probabilities of events depend upon the causal relations into which they enter.

Third, causal relations have some important formal properties. It would seem, for example, to be an essential property of causation that it is asymmetric, and that causal loops are impossible. One test, therefore, of the adequacy of any proposed account of causation is whether it follows from that account that causal relations have the formal properties they are normally taken to have. Moreover, if it does follow, one also wants to ask if this is a consequence of the core idea upon which the account is based, rather than flowing instead from some additional, *ad hoc* postulates.

Much of the discussion in Chapter 8 was directed to this issue. I attempted to show that it does follow from the analyses offered that causal relations possess the appropriate formal properties, and that this is a consequence simply of postulates that are needed to capture, in a precise way, the intuitive idea that causal relations are relations that determine the direction of the logical transmission of probabilities.

Fourth, another important test for any account of causation is whether it is possible, given that account, to have knowledge of causal relations, and whether the sorts of considerations that one normally takes as providing evidence of the presence of causal connections do so on the account in question. This issue was discussed in section 9.2, where it was argued that the realist account of causation which has been proposed here is epistemologically adequate.

In addition to these primary issues, two secondary topics were discussed at some length. First, there was the question of a singularist conception of causation. Second, there was the question of the relation between causation and time.

With regard to the former, two main points were made. In Chapter 6, I tried to show that, contrary to what is generally believed, there are some quite strong arguments against a supervenience view of causal relations—arguments that definitely need to be answered if one is to embrace such an account. Then, in Chapter 8, the crucial question of whether one can, in the final analysis, make sense of a singularist conception of causation was addressed, and I tried to show that there are serious difficulties that appear to stand in the way of doing so.

As I do not believe, either that there are satisfactory answers to the objections to a supervenience account set out in Chapter 6, or that one can make sense of a singularist conception, the approach to causation which now seems to me to be correct is one which is intermediate between supervenience and singularist views. But as I stressed above, and as will be clear from the discussion in

Chapter 8, my general approach to causation is perfectly compatible with a supervenience account.

As regards the question of the relation between time and causation, I argued, in Chapter 7, that the direction of causation is not to be explained in terms of the direction of time. I then went on, in Chapter 8, to set out an account of causation which does not involve any temporal concepts. In contrast to what is the case on many accounts of causation, therefore, the possibility of a causal theory of time is not precluded.

This possibility was briefly discussed in section 9.3. There I tried to show that it was important, in evaluating the prospects for a causal theory of time, not to assume either that such a theory is committed to a relational conception of space and time, or that it must be combined with a tenseless view of the nature of time. A causal theory of time is, as I argued there, perfectly compatible with a substantival view of space and time, and with a tensed approach to time. I suggested, moreover, that the most promising form for a causal theory of time appears to be one which involves both a tensed account of time, and a substantival view of space and time. For, on the one hand, the combination of a causal theory of time and a substantival view of space and time enables one to avoid some quite serious objections to which relational theories are exposed; and, on the other, a causal theory of time which does treat tense as ontologically significant enables one, I believe, to capture more fully the idea that causation involves the bringing into existence of one event by another.

The views that I have advanced here—both in the case of laws, and in the case of causation—have aspects that, I must confess, sometimes seem to me less than completely unproblematic. But I am convinced that there is a very strong case for a realist account, both of the nature of laws, and of causation, and I am hopeful that the specific analyses that I have proposed are along roughly the right lines. As for any errors that remain, I hope that the underlying issues have at least been formulated in a way which will facilitate further discussion of these central metaphysical questions.

APPENDIX

Proofs of two theorems, from section 8.4.2, are given below—
Theorem 15, and Theorem 19.

THEOREM 15: CAUSAL NECESSITATION IS RELATED TO INCREASE IN PROBABILITY

In the case of the relation of direct causal necessitation, it was
shown, in Theorem 7, that the existence of a basic causal law
connecting two properties is related to an increase in the
likelihood of instantiation of one of those properties. In particular,
it was established that, provided S satisfies requirement T,
$\text{Prob}(Qx, P \rightarrow_k Q \,\&\, S)$ is always at least equal to $\text{Prob}(Qx, S)$, and
is greater than $\text{Prob}(Qx, S)$, as long as it is not the case either that
$\text{Prob}(Qx, \sim Px \,\&\, S)$ is equal to one, or that $\text{Prob}(Px, S)$ is equal to
zero.

A comparable result can be established for the ancestral. The
first step involves showing that:

(1) $\text{Prob}(Qx, P_n \rightarrow_k P_{n-1} \,\&\, P_{n-1} \rightarrow_k P_{n-2} \,\&\, \ldots \,\&\, P_1 \rightarrow_k Q \,\&\, S) =$
$\text{Prob}(P_n x, S) + \text{Prob}(\sim P_n x \,\&\, P_{n-1} x, S) + \ldots +$
$\text{Prob}(\sim P_n x \,\&\, \sim P_{n-1} x \,\&\, \ldots \sim P_2 x \,\&\, P_1 x, S) +$
$\text{Prob}(\sim P_n x \,\&\, \sim P_{n-1} x \,\&\, \ldots \,\&\, \sim P_2 x \,\&\, \sim P_1 x, S) \times$
$\text{Prob}(Qx, \sim P_n x \,\&\, \sim P_{n-1} x \,\&\, \ldots \sim P_2 x \,\&\, \sim P_1 x \,\&\, S)$,
provided that S satisfies requirement T.

This can be established by mathematical induction. First, if n is
equal to one, the above reduces to

$\text{Prob}(Qx, P_1 \rightarrow_k Q \,\&\, S) =$
$\text{Prob}(P_1 x, S) + \text{Prob}(\sim P_1 x, S) \times \text{Prob}(Qx, \sim P_1 x \,\&\, S)$,
provided that S satisfies requirement T,

and this is just a case of postulate (R_5).

Next, we need to show that if equation (1) is true for some n, it
will also be true for $(n + 1)$. The argument here will be slightly
more perspicuous if we abbreviate the expression '$P_n \rightarrow_k P_{n-1} \,\&\,$

$P_{n-1} \to_k P_{n-2}$ & . . . & $P_1 \to_k Q$ & S' to 'R'. Given this abbreviation, what needs to be worked out is the value of $\text{Prob}(Qx, P_{n+1} \to_k P_n \ \& \ R)$. This can be done by using the following truth of probability theory:

$$\text{Prob}(B, A) = \text{Prob}(B, C \ \& \ A) \times \text{Prob}(C, A) + \text{Prob}(B, {\sim}C \ \& \ A) \\ \times \text{Prob}({\sim}C, A).$$

Applying this in the present case gives us that:

(2) $\text{Prob}(Qx, P_{n+1} \to_k P_n \ \& \ R) =$
$\text{Prob}(Qx, P_{n+1}x \ \& \ P_{n+1} \to_k P_n \ \& \ R) \times \text{Prob}(P_{n+1}x, P_{n+1} \to_k P_n \ \& \ R) +$
$\text{Prob}(Qx, {\sim}P_{n+1}x \ \& \ P_{n+1} \to_k P_n \ \& \ R) \times \text{Prob}({\sim}P_{n+1}x, P_{n+1} \to_k P_n \ \& \ R)$.

But if S satisfies requirement T, we have that:

(3) $\text{Prob}(Qx, P_{n+1}x \ \& \ P_{n+1} \to_k P_n \ \& \ R) = 1$,

by virtue of postulate (R_6), given what R is;

(4) $\text{Prob}(P_{n+1}x, P_{n+1} \to_k P_n \ \& \ R) = \text{Prob}(P_{n+1}x, S)$,

by virtue of postulate (R_2), given what R is;

(5) $\text{Prob}(Qx, {\sim}P_{n+1} \ \& \ P_{n+1} \to_k P_n \ \& \ R) =$
$\text{Prob}(Qx, {\sim}P_{n+1}x \ \& \ R)$,

by postulate (R_4);

(6) $\text{Prob}({\sim}P_{n+1}x, P_{n+1} \to_k P_n \ \& \ R) = \text{Prob}({\sim}P_{n+1}x, S)$,

by virtue of postulate (R_2), an elementary truth of probability theory, and the definition of R.

When substitutions are made in equation (2), above, using equations (3) to (6), one has that

(7) $\text{Prob}(Qx, P_{n+1} \to_k P_n \ \& \ R) =$
$\text{Prob}(P_{n+1}x, S) + \text{Prob}({\sim}P_{n+1}x, S) \times \text{Prob}(Qx, {\sim}P_{n+1}x \ \& \ R)$.

We now need an equation that gives us the value of $\text{Prob}(Qx, {\sim}P_{n+1}x \ \& \ R)$, i.e., of $\text{Prob}(Qx, {\sim}P_{n+1}x \ \& \ P_n \to_k P_{n-1} \ \& \ P_{n-1} \to_k P_{n-2} \ \& \ . . . \ \& \ P_2 \to_k P_1 \ \& \ P_1 \to_k Q \ \& \ S)$. Since we are assuming that equation (1) holds for n, and are trying to show that it then follows that it holds for $(n + 1)$, we can, in equation (1), replace S by ${\sim}P_{n+1}x \ \& \ S$. When this substitution is made we have:

(8) $\text{Prob}(Qx, \sim P_{n+1}x \ \& \ P_n \rightarrow_k P_{n-1} \ \& \ P_{n-1} \rightarrow_k P_{n-2} \ \& \ \dots \ \& \ P_1 \rightarrow_k Q \ \& \ S) =$
$\text{Prob}(P_n x, \sim P_{n+1}x \ \& \ S) + \text{Prob}(\sim P_n x \ \& \ P_{n-1}x, \sim P_{n+1}x \ \& \ S) + \dots +$
$\text{Prob}(\sim P_n x \ \& \ \sim P_{n-1}x \ \& \ \dots \ \sim P_2 x \ \& \ P_1 x, \sim P_{n+1}x \ \& \ S) +$
$\text{Prob}(\sim P_n x \ \& \ \sim P_{n-1}x \ \& \ \dots \ \& \ \sim P_2 x \ \& \ \sim P_1 x, \sim P_{n+1}x \ \& \ S) \times$
$\text{Prob}(Qx, \sim P_{n+1}x \ \& \ \sim P_n x \ \& \ \sim P_{n-1}x \ \& \ \dots \ \sim P_2 x \ \& \ \sim P_1 x \ \& \ S).$

When this equality is substituted in equation (7), and the products are simplified, using the fact that

$$\text{Prob}(A \ \& \ B, C) = \text{Prob}(A, B \ \& \ C) \times \text{Prob}(B, C)$$

we have that

(9) $\text{Prob}(Qx, P_{n+1} \rightarrow_k P_n \ \& \ P_n \rightarrow_k P_{n-1} \ \& \ \dots \ \& \ P_1 \rightarrow_k Q \ \& \ S) =$
$\text{Prob}(P_{n+1}x, S) + \text{Prob}(\sim P_{n+1}x \ \& \ P_n x, S) + \dots +$
$\text{Prob}(\sim P_{n+1}x \ \& \ \sim P_n x \ \& \ \sim P_{n-1}x \ \& \ \dots \ \& \ \sim P_2 x \ \& \ P_1 x, S) +$
$\text{Prob}(\sim P_{n+1}x \ \& \ \sim P_n x \ \& \ \sim P_{n-1}x \ \& \ \dots \ \& \ \sim P_2 x \ \& \ \sim P_1 x, S) \times$
$\text{Prob}(Qx, \sim P_{n+1}x \ \& \ \sim P_n x \ \& \ \sim P_{n-1}x \ \& \ \dots \ \& \ \sim P_2 x \ \& \ \sim P_1 x \ \& \ S).$

This shows that if the equality holds for some n, it must also hold for $(n + 1)$. This completes the inductive step, and together with the fact that the equality does hold for n equal to one, shows that it holds for any n.

The final step in the argument involves comparing the equation just established with an appropriate equation for $\text{Prob}(Qx, S)$, namely:

(10) $\text{Prob}(Qx, S) =$
$\text{Prob}(Qx, P_n x \ \& \ S) \times \text{Prob}(P_n x, S) +$
$\text{Prob}(Qx, \sim P_n x \ \& \ P_{n-1}x \ \& \ S) \times \text{Prob}(\sim P_n x \ \& \ P_{n-1}x, S) + \dots +$
$\text{Prob}(Qx, \sim P_n x \ \& \ \sim P_{n-1}x \ \& \ \dots \ \sim P_2 x \ \& \ P_1 x \ \& \ S) \times$
$\text{Prob}(\sim P_n x \ \& \ \sim P_{n-1}x \ \& \ \dots \ \& \ \sim P_2 x \ \& \ P_1 x, S) +$
$\text{Prob}(Qx, \sim P_n x \ \& \ \sim P_{n-1}x \ \& \ \dots \ \sim P_2 x \ \& \ \sim P_1 x \ \& \ S) \times$
$\text{Prob}(\sim P_n x \ \& \ \sim P_{n-1}x \ \& \ \dots \ \& \ \sim P_2 x \ \& \ \sim P_1 x, S).$

It can be readily seen, by comparing equations (1) and (10), that, subject to the condition that S satisfies requirement T, Prob(Qx, $P_n \to_k P_{n-1}$ & $P_{n-1} \to_k P_{n-2}$ & . . . & $P_1 \to_k Q$, S) will be greater than Prob(Qx, S) unless it is true, not only that either Prob(Qx, $P_n x$ & S) is equal to one, or Prob($P_n x$, S) is equal to zero, but also that, for all j from zero to $(n - 2)$, either Prob(Qx, $\sim P_n x$ & $\sim P_{n-1} x$ & . . . $\sim P_{n-j} x$ & $P_{n-j-1} x$ & S) is equal to one, or Prob($\sim P_n x$ & $\sim P_{n-1} x$ & . . . & $\sim P_{n-j} x$ & $P_{n-j-1} x$, S) is equal to zero.

To sum up, then, if S satisfies requirement T, Prob(Qx, $P \to^* Q$ & S) will never be less than Prob(Qx, S), and it is only in quite exceptional circumstances that it will fail to be greater.

THEOREM 19: CAUSAL LOOPS ARE IMPOSSIBLE

One has a causal loop if there are properties (or quasi-properties) P, P_1, Q_1, P_2, Q_2, . . . P_n, Q_n, such that either of the following is true:

(1) $P \to^* Q_1$ & $P_1 \to^* Q_2$ & . . . & $P_{n-1} \to^* Q_n$ & $P_n \to^* P$;
(2) $P \to^* Q_1$ & $P_1 \to^* Q_2$ & . . . & $P_{n-1} \to^* Q_n$ & $P_n \to^* \sim P$,

where, for every i, P_i is either identical with Q_i, or with $\sim Q_i$. The present theorem shows that neither of these can be the case.

In order to prove this, we need to establish the following:

Prob(Px, $P \to^* Q_1$ & $P_1 \to^* Q_2$ & . . . & $P_{n-1} \to^* Q_n$ & $P_n \to^* R$ & S)

must be equal either to

Prob(Px, $P \to^* P_n$ & $P_n \to^* R$ & S),

or else to

Prob(Px, $P \to^* \sim P_n$ & $P_n \to^* R$ & S).

To establish this, it will be helpful to prove the following four lemmas:

Lemma 1: Prob(Px, $P \to^* Q$ & S) = Prob(Px, $\sim P \to^* \sim Q$ & S), provided that S satisfies requirement T;
Lemma 2: Prob(Px, $P \to^* Q$ & $\sim Q \to^* R$ & S) = Prob(Px, S), provided that S satisfies requirement T;

Lemma 3: $\text{Prob}(Px, P \rightarrow^* Q \ \& \ Q \rightarrow^* R \ \& \ S) = \text{Prob}(Px, P \rightarrow^* R \ \& \ S)$,
      provided that S satisfies requirement T;

Lemma 4: $\text{Prob}(Px, \sim P \rightarrow^* \sim Q \ \& \ \sim Q \rightarrow^* \sim R \ \& \ S) = \text{Prob}(Px, \sim P \rightarrow^* \sim R \ \& \ S)$,
      provided that S satisfies requirement T.

Lemma 1: $\text{Prob}(Px, P \rightarrow^* Q \ \& \ S) = \text{Prob}(Px, \sim P \rightarrow^* \sim Q \ \& \ S)$,
      provided that S satisfies requirement T.

By virtue of postulate $(R_1{}^*)$, we have that

$$\text{Prob}(Px, P \rightarrow^* Q \ \& \ S) = \text{Prob}(Px, S),$$

and also that

$$\text{Prob}(\sim Px, \sim P \rightarrow^* \sim Q \ \& \ S) = \text{Prob}(\sim Px, S).$$

It follows from the latter, by probability theory, that

$$\text{Prob}(Px, \sim P \rightarrow^* \sim Q \ \& \ S) = \text{Prob}(Px, S).$$

Therefore we have that

$$\text{Prob}(Px, P \rightarrow^* Q \ \& \ S) = \text{Prob}(Px, \sim P \rightarrow^* \sim Q \ \& \ S).$$

Lemma 2: $\text{Prob}(Px, P \rightarrow^* Q \ \& \ \sim Q \rightarrow^* R \ \& \ S) = \text{Prob}(Px, S)$,
      provided that S satisfies requirement T.

By virtue of lemma 1,

$$\text{Prob}(Px, P \rightarrow^* Q \ \& \ \sim Q \rightarrow^* R \ \& \ S) =$$
$$\text{Prob}(Px, \sim P \rightarrow^* \sim Q \ \& \ \sim Q \rightarrow^* R \ \& \ S).$$

This in turn is equal to

$$1 - \text{Prob}(\sim Px, \sim P \rightarrow^* \sim Q \ \& \ \sim Q \rightarrow^* R \ \& \ S),$$

which, by virtue of postulate $(R_2{}^*)$, is equal to

$$1 - \text{Prob}(\sim Px, S),$$

and thus to

$$\text{Prob}(Px, S).$$

Therefore we have that

$$\text{Prob}(Px, P \rightarrow^* Q \ \& \ \sim Q \rightarrow^* R \ \& \ S) = \text{Prob}(Px, S).$$

Lemma 3: $\text{Prob}(Px, P \to^* Q \,\&\, Q \to^* Rx \,\&\, S) = \text{Prob}(Px, P \to^* R \,\&\, S)$,
   provided that S satisfies requirement T.

By virtue of postulates (R_1^*) and (R_2^*) we have, respectively, that

$$\text{Prob}(Px, P \to^* R \,\&\, S) = \text{Prob}(Px, S),$$

and that

$$\text{Prob}(Px, P \to^* Q \,\&\, Q \to^* R \,\&\, S) = \text{Prob}(Px, S).$$

Therefore

$$\text{Prob}(Px, P \to^* Q \,\&\, Q \to^* R \,\&\, S) = \text{Prob}(Px, P \to^* R \,\&\, S).$$

Lemma 4: $\text{Prob}(Px, \sim P \to^* \sim Q \,\&\, \sim Q \to^* \sim R \,\&\, S) = \text{Prob}(Px, \sim P \to^* \sim R \,\&\, S)$,
   provided that S satisfies requirement T.

By virtue of lemma 3, we have the following:

$$\text{Prob}(\sim Px, \sim P \to^* \sim Q \,\&\, \sim Q \to^* \sim R \,\&\, S) = \text{Prob}(\sim Px, \sim P \to^* \sim R \,\&\, S).$$

It then follows immediately that

$$\text{Prob}(Px, \sim P \to^* \sim Q \,\&\, \sim Q \to^* \sim R \,\&\, S) = \text{Prob}(Px, \sim P \to^* \sim R \,\&\, S).$$

Given these four lemmas, the proof of the theorem now runs as follows. Consider

$$\text{Prob}(Px, P \to^* Q_1 \,\&\, P_1 \to^* Q_2 \,\&\, \ldots \,\&\, P_n \to^* R \,\&\, S).$$

By virtue of the constraint upon the P_i, either P_1 is identical with Q_1 or it is identical with $\sim Q_1$. Suppose that P_1 is identical with Q_1. Then lemma 3 can be applied, and we shall have that

$$\text{Prob}(Px, P \to^* Q_1 \,\&\, P_1 \to^* Q_2 \,\&\, \ldots \,\&\, P_n \to^* R \,\&\, S) =$$
$$\text{Prob}(Px, P \to^* Q_2 \,\&\, P_2 \to^* Q_3 \,\&\, \ldots \,\&\, P_n \to^* R \,\&\, S)$$

That is to say, the first two conjuncts, $P \to^* Q_1$ and $P_1 \to^* Q_2$, can be replaced by a single conjunct, $P \to^* Q_2$. Suppose, on the other hand, that P_1 is identical with $\sim Q_1$. Then we need to make use of lemma 1 to insert negation signs in front of the two formulas in the first conjunct. Thus we have that

Prob(Px, $P \rightarrow^* Q_1$ & $P_1 \rightarrow^* Q_2$ & . . . & $P_n \rightarrow^* R$ & S) =
Prob(Px, $\sim P \rightarrow^* \sim Q_1$ & $P_1 \rightarrow^* Q_2$ & . . . & $P_n \rightarrow^* R$ & S).

But by virtue of lemma 4, plus the hypothesis that P_1 is identical with $\sim Q_1$, we have that

· Prob(Px, $\sim P \rightarrow^* \sim Q_1$ & $P_1 \rightarrow^* Q_2$ & . . . & $P_n \rightarrow^* R$ & S) =
Prob(Px, $\sim P \rightarrow^* Q_2$ & $P_2 \rightarrow^* Q_3$ & . . . & $P_n \rightarrow^* R$ & S).

Combining these we then have that

Prob(Px, $P \rightarrow^* Q_1$ & $P_1 \rightarrow^* Q_2$ & . . . & $P_n \rightarrow^* R$ & S) =
Prob(Px, $\sim P \rightarrow^* Q_2$ & $P_2 \rightarrow^* Q_3$ & . . . & $P_n \rightarrow^* R$ & S).

We can then use lemma 1 again, to insert negation signs in front of the two formulas in the first conjunct, i.e., in $\sim P \rightarrow^* Q_2$; and when this is done, and the resulting double negation signs are dropped, we have that

Prob(Px, $P \rightarrow^* Q_1$ & $P_1 \rightarrow^* Q_2$ & . . . & $P_n \rightarrow^* R$ & S) =
Prob(Px, $P \rightarrow^* \sim Q_2$ & $P_2 \rightarrow^* Q_3$ & . . . & $P_n \rightarrow^* R$ & S).

This shows that, regardless of whether P_1 is identical with Q_1 or identical with $\sim Q_1$, the first two conjuncts, $P \rightarrow^* Q_1$ and $P_1 \rightarrow^* Q_2$, can be replaced by a single conjunct, which will either be $P \rightarrow^* Q_2$, or $P \rightarrow^* \sim Q_2$.

Since P_2 is either identical with Q_2 or with $\sim Q_2$, the process can be repeated. By carrying it out $(n-1)$ times, one can show that

Prob(Px, $P \rightarrow^* Q_1$ & $P_1 \rightarrow^* Q_2$ & . . . & $P_n \rightarrow^* R$ & S)

must be equal either to

Prob(Px, $P \rightarrow^* Q_n$ & $P_n \rightarrow^* R$ & S),

or else to

Prob(Px, $P \rightarrow^* \sim Q_n$ & $P_n \rightarrow^* R$ & S).

Recall, now, what needs to be shown; namely, that there cannot be properties (or quasi-properties) P, P_1, Q_1, P_2, Q_2, . . . P_n, Q_n, such that at least one of the following is true:

(1) $P \rightarrow^* Q_1$ & $P_1 \rightarrow^* Q_2$ & . . . & $P_{n-1} \rightarrow^* Q_n$ & $P_n \rightarrow^* P$;
(2) $P \rightarrow^* Q_1$ & $P_1 \rightarrow^* Q_2$ & . . . & $P_{n-1} \rightarrow^* Q_n$ & $P_n \rightarrow^* \sim P$

where for every i, either P_i is identical with Q_i or with $\sim Q_i$. If it is possible for (1) to be true, then the following probability must be defined:

$$\text{Prob}(Px, P \to^* Q_1 \,\&\, P_1 \to^* Q_2 \,\&\, \ldots \,\&\, P_{n-1} \to^* Q_n \,\&\, P_n \to^* P \,\&\, S).$$

And by what has just been established, this must be equal either to

$$\text{Prob}(Px, P \to^* Q_n \,\&\, P_n \to^* P \,\&\, S),$$

or else to

$$\text{Prob}(Px, P \to^* {\sim}Q_n \,\&\, P_n \to^* P \,\&\, S).$$

But if either of these were defined, a contradiction would arise. This can be seen as follows. First, consider

$$\text{Prob}(Px, P \to^* Q_n \,\&\, P_n \to^* P \,\&\, S).$$

By postulate $(R_1{}^*)$, this must be equal to

$$\text{Prob}(Px, P_n \to^* P \,\&\, S).$$

If P_n is identical with Q_n, we can also apply postulate $(R_2{}^*)$, which will give us that

$$\text{Prob}(Px, P \to^* Q_n \,\&\, P_n \to^* P \,\&\, S) = \text{Prob}(Px, S).$$

If, on the other hand, P_n is identical with ${\sim}Q_n$, we can apply lemma 2, which will lead to the same result. Accordingly, it follows that

$$\text{Prob}(Px, P_n \to^* P \,\&\, S) = \text{Prob}(Px, S).$$

In Theorem 15, it was shown, however, that if $\text{Prob}(Qx, Px \,\&\, S)$ is less than one, and $\text{Prob}(Px, S)$ is greater than zero, then $\text{Prob}(Qx, P \to^* Q \,\&\, S)$ is greater than $\text{Prob}(Qx, S)$. It must therefore be the case that, provided $\text{Prob}(Px, P_n x \,\&\, S)$ is less than one, and $\text{Prob}(P_n x, S)$ is greater than zero, $\text{Prob}(Px, P_n \to^* P \,\&\, S)$ is greater than $\text{Prob}(Px, S)$. That there must exist an S satisfying requirement T which is such that $\text{Prob}(Px, P_n x \,\&\, S)$ is less than one, and $\text{Prob}(P_n x, S)$ is greater than zero, can be justified in the same way as similar conditions in earlier arguments. So there will be an S, satisfying requirement T, such that

$$\text{Prob}(Px, P_n \to^* P \,\&\, S) \text{ is greater than } \text{Prob}(Px, S),$$

and this, together with

$\text{Prob}(Px, P_n \rightarrow^* P \, \& \, S) = \text{Prob}(Px, S)$,

entails a contradiction. Next, consider

$\text{Prob}(Px, P \rightarrow^* \sim Q_n \, \& \, P_n \rightarrow^* P \, \& \, S)$.

If P_n is identical with $\sim Q_n$, then this will be equal to $\text{Prob}(Px, S)$, by virtue of postulate $(R_2{}^*)$. If, on the other hand, P_n is identical with Q_n, we can first apply lemma I, which will give us that

$\text{Prob}(Px, P \rightarrow^* \sim Q_n \, \& \, P_n \rightarrow^* P \, \& \, S) = \text{Prob}(Px, \sim P \rightarrow^* Q_n \, \& \, P_n \rightarrow^* P \, \& \, S)$.

The latter, in turn, is equal to

$\text{I} - \text{Prob}(\sim Px, \sim P \rightarrow^* Q_n \, \& \, P_n \rightarrow^* P \, \& \, S)$.

Then, since P_n is identical with Q_n, postulate $(R_2{}^*)$ can be applied, with the result that this must be equal to

$\text{I} - \text{Prob}(\sim Px, S)$,

and so equal to

$\text{Prob}(Px, S)$.

Therefore, regardless of whether P_n is identical with Q_n or with $\sim Q_n$, it must be the case that

$\text{Prob}(Px, P \rightarrow^* \sim Q_n \, \& \, P_n \rightarrow^* P \, \& \, S) = \text{Prob}(Px, S)$.

But by postulate $(R_1{}^*)$, we also have that

$\text{Prob}(Px, P \rightarrow^* \sim Q_n \, \& \, P_n \rightarrow^* P \, \& \, S) = \text{Prob}(Px, P_n \rightarrow^* P \, \& \, S)$.

Therefore

$\text{Prob}(Px, P_n \rightarrow^* P \, \& \, S) = \text{Prob}(Px, S)$.

And this, by virtue of the fact that there must be an S satisfying requirement T such that $\text{Prob}(Px, P_n \rightarrow^* P \, \& \, S)$ is greater than $\text{Prob}(Px, S)$, leads to the same contradiction as in the preceding case.

Consequently, the following cannot be true:

(1) $P \rightarrow^* Q_1 \, \& \, P_1 \rightarrow^* Q_2 \, \& \, \ldots P_{n-1} \rightarrow^* Q_n \, \& \, P_n \rightarrow^* P$.

This leaves us with the possibility that

(2) $P \rightarrow^* Q_1$ & $P_1 \rightarrow^* Q_2$ & . . . $P_{n-1} \rightarrow^* Q_n$ & $P_n \rightarrow^* \sim P$.

But this can be ruled out by a similar line of argument. For if (2) were true, then the following probability would be defined:

$\text{Prob}(Px, P \rightarrow^* Q_1$ & $P_1 \rightarrow^* Q_2$ & . . . & $P_{n-1} \rightarrow^* Q_n$ & $P_n \rightarrow^* \sim P$ & $S)$.

By the result established above, this must be equal either to

$\text{Prob}(Px, P \rightarrow^* Q_n$ & $P_n \rightarrow^* \sim P$ & $S)$,

or else to

$\text{Prob}(Px, P \rightarrow^* \sim Q_n$ & $P_n \rightarrow^* \sim P$ & $S)$.

First, consider

$\text{Prob}(Px, P \rightarrow^* Q_n$ & $P_n \rightarrow^* \sim P$ & $S)$.

By postulate (R_1^*), this must be equal to

$\text{Prob}(Px, P_n \rightarrow^* \sim P$ & $S)$.

If P_n is identical with Q_n, then postulate (R_2^*) implies that it must also be equal to

$\text{Prob}(Px, S)$.

If, on the other hand, P_n is identical, instead, with $\sim Q_n$, then the same result follows from lemma 2. So in either case we have that

$\text{Prob}(Px, P_n \rightarrow^* \sim P$ & $S) = \text{Prob}(Px, S)$.

But, given Theorem 15, it must be the case that $\text{Prob}(\sim Px, P_n \rightarrow^* \sim P$ & $S)$ is greater than $\text{Prob}(\sim Px, S)$, provided that there is an S satisfying requirement T, such that $\text{Prob}(\sim Px, P_n x$ & $S)$ is less than one, and $\text{Prob}(P_n x, S)$ is greater than zero, from which it follows, by probability theory, that $\text{Prob}(Px, P_n \rightarrow^* \sim P$ & $S)$ is less than $\text{Prob}(Px, S)$. Since those conditions can be satisfied, we once again have a contradiction.

Finally, consider

$\text{Prob}(Px, P \rightarrow^* \sim Q_n$ & $P_n \rightarrow^* \sim P$ & $S)$.

If P_n is equal to Q_n, we can apply lemma 1, getting that

$\text{Prob}(Px, P \rightarrow^* \sim Q_n$ & $P_n \rightarrow^* \sim P$ & $S) = $
$\text{Prob}(Px, \sim P \rightarrow^* Q_n$ & $P_n \rightarrow^* \sim P$ & $S)$.

This, in turn, is equal to

$$1 - \text{Prob}(\sim Px, \sim P \to^* Q_n \ \& \ P_n \to^* \sim P \ \& \ S).$$

Then, since we are considering the case where P_n is identical with Q_n, it follows by virtue of postulate $(R_2{}^*)$ that this is equal to

$$1 - \text{Prob}(\sim Px, S)$$

and thus to

$$\text{Prob}(Px, S).$$

Therefore, if P_n is identical with Q_n, we have that

$$\text{Prob}(Px, P \to^* \sim Q_n \ \& \ P_n \to^* \sim P \ \& \ S) = \text{Prob}(Px, S).$$

On the other hand, if P_n is identical, instead, with $\sim Q_n$, then the same result follows immediately by virtue of postulate $(R_2{}^*)$. But by postulate $(R_1{}^*)$, we also have that

$$\text{Prob}(Px, P \to^* \sim Q_n \ \& \ P_n \to^* \sim P \ \& \ S) = \text{Prob}(Px, P_n \to^* \sim P \ \& \ S).$$

It is therefore the case that

$$\text{Prob}(Px, P_n \to^* \sim P \ \& \ S) = \text{Prob}(Px, S).$$

And this, by virtue of the fact that there must be an S, satisfying requirement T, such that $\text{Prob}(Px, P_n \to^* \sim P \ \& \ S)$ is less than $\text{Prob}(Px, S)$, leads to the same contradiction as in the preceding case.

It has now been established that there cannot be properties P, $P_1, Q_1, P_2, Q_2 \ldots P_n$, and Q_n such that either of the following is the case:

(1) $P \to^* Q_1 \ \& \ P_1 \to^* Q_2 \ \& \ldots \& \ P_{n-1} \to^* Q_n \ \& \ P_n \to^* P$;
(2) $P \to^* Q_1 \ \& \ P_1 \to^* Q_2 \ \& \ldots \& \ P_{n-1} \to^* Q_n \ \& \ P_n \to^* \sim P$

where for every i, P_i is identical either with Q_i or with $\sim Q_i$. Causal loops, involving the ancestral of the relation of direct causal necessitation, are therefore impossible.

BIBLIOGRAPHY

ANSCOMBE, G. ELIZABETH M., 'Causality and Determination', *Causation and Conditionals*, ed. E. Sosa, (Oxford, 1975), 63–81.
ARMSTRONG, DAVID M., *A Materialist Theory of the Mind* (New York, 1968).
—— *Universals and Scientific Realism*, 2 vols., (Cambridge, 1978).
—— *What Is a Law of Nature?* (Cambridge, 1983).
BAR-HILLEL, YEHOSHUA (ed.), *Logic, Methodology and Philosophy of Science* (Amsterdam, 1965).
BEAUCHAMP, TOM L. (ed.), *Philosophical Problems in Causation* (Encino and Belmont, Calif., 1974).
—— AND ROSENBERG, ALEXANDER, *Hume and the Problem of Causation* (New York and Oxford, 1981).
BENNETT, JONATHAN, 'Counterfactuals and Possible Worlds', *Canadian Journal of Philosophy*, 4 (1974), 391–402.
BLACK, MAX, 'Why Cannot an Effect Precede its Cause?', *Analysis*, 16 (1955–6), 49–58.
BOLTZMANN, LUDWIG, Vorlesung über die kinetische Theorie der Gase (Leipzig, 1902).
BRAND, MYLES, 'Simultaneous Causation', *Time and Cause*, ed. P. van Inwagen (Dordrecht, 1980), 137–53.
BRETZEL, PHILIP VON, 'Concerning a Probabilistic Theory of Causation Adequate for the Causal Theory of Time', *Synthese*, 35 (1977), 173–90.
CARNAP, RUDOLF, *The Logical Foundations of Probability*, 2nd edn., (Chicago, 1962).
CHISHOLM, RODERICK M., 'Law Statements and Counterfactual Inference', *Analysis*, 15 (1955), 97–105.
DEVITT, MICHAEL, ' "Ostrich Nominalism" or "Mirage Realism"?', *Pacific Philosophical Quarterly*, 61 (1980), 433–9.
DRETSKE, FRED I., 'Laws of Nature', *Philosophy of Science*, 44 (1977), 248–68.
DUCASSE, C. J., 'On the Nature and the Observability of the Causal Relation', *Journal of Philosophy*, 23 (1926), 57–67, and reprinted in *Causation and Conditionals*, ed. E. Sosa (Oxford, 1975).
DUMMETT, MICHAEL, 'Can an Effect Precede its Cause?', *Proceedings of the Aristotelian Society*, Suppl. 28 (1954), 27–44.
—— 'Bringing About the Past', *Philosophical Review*, 73 (1964), 338–59.
EHRENFEST, P. AND T., 'Begriffliche Grundlagen der statistischen Auffassung in der Mechanik', Encyclopädie der mathematischen Wissenschaften, 4, 2, II (Leipzig, 1911), 41–51.

ENGLISH, JANE, 'Underdetermination: Craig and Ramsey', *Journal of Philosophy*, 70 (1973), 453–63.

FEIGL, HERBERT AND MAXWELL, GROVER (eds.), *Current Issues in the Philosophy of Science* (New York, 1961).

FIELD, HARTRY, *Science Without Numbers* (Princeton, 1980).

FISK, MILTON, 'Are There Necessary Connections in Nature?', *Philosophy of Science*, 37/3 (1970), 385–404.

FLEW, ANTONY, 'Can an Effect Precede its Cause?', *Proceedings of the Aristotelian Society*, Suppl. 28 (1954), 45–62.

—— 'Effects before their Causes—Addenda and Corrigenda', *Analysis*, 16 (1955–6), 104–10.

—— 'Causal Disorder Again', *Analysis*, 17 (1956–7), 81–6.

FOSTER, JOHN, 'Induction, Explanation and Natural Necessity', *The Aristotelian Society*, 83 (1982–3), 87–101.

FREEMAN, E. AND SELLARS, W. (eds.), *Basic Issues in the Philosophy of Time* (La Salle, Ill., 1971).

GALE, R. M., 'Why a Cause Cannot be Later than its Effect', *Review of Metaphysics*, 19 (1965), 209–34.

GASKING, DOUGLAS, 'Causation and Recipes', *Mind*, 64/256 (1955), 479–87.

GIBBS, J. W., *Elementary Principles in Statistical Mechanics* (New York, 1922).

GOROVITZ, SAMUEL, 'Leaving the Past Alone', *Philosophical Review*, 73 (1964), 360–71.

GRÜNBAUM, ADOLF, 'Carnap's Views on the Foundations of Geometry', *The Philosophy of Rudolf Carnap*, ed. P. A. Schilpp (La Salle, Ill., 1963), 599–684.

—— *Philosophical Problems of Space and Time*, 2nd edn., (Dordrecht, 1973).

HEMPEL, C. G., 'Studies in the Logic of Confirmation', *Mind*, 54 (1945), 1–26 and 97–121.

—— *Aspects of Scientific Explanation* (New York, 1965).

HETHERINGTON, STEPHEN C., 'Tooley's Theory of Laws of Nature', *Canadian Journal of Philosophy*, 13/1 (1983), 101–5.

HILBERT, DAVID, *Foundations of Geometry* (La Salle, Ill., 1971).

HUME, DAVID, *A Treatise of Human Nature*, i and ii (London, 1739), iii (London, 1740).

—— *An Inquiry Concerning Human Understanding* (London, 1748).

INWAGEN, PETER VAN (ed.), *Time and Cause* (Dordrecht, 1980).

JACKSON, FRANK, 'A Causal Theory of Counterfactuals', *Australasian Journal of Philosophy*, 55 (1977), 3–21.

—— 'Statements About Universals', *Mind*, 86 (1977), 427–9.

JEFFREY, RICHARD C. (ed.), *Studies in Inductive Logic and Probability* (Berkeley and Los Angeles, Calif., 1980).

KNEALE, WILLIAM, *Probability and Induction* (Oxford, 1949).
—— 'Natural Laws and Contrary-to-Fact Conditionals', *Analysis*, 10/6 (1950), 121–5.
—— 'Universality and Necessity', *British Journal for the Philosophy of Science*, 12/46 (1961), 89–102.
KORNER, STEPHAN (ed.), *Explanation* (Oxford, 1975).
LEWIS, DAVID, 'How to Define Theoretical Terms', *Journal of Philosophy*, 67 (1970), 427–46.
—— *Counterfactuals* (Cambridge, Mass., 1973).
—— 'A Subjectivist's Guide to Objective Chance', *Studies in Inductive Logic and Probability*, ii, ed. R. C. Jeffrey (Berkeley and Los Angeles, Calif., 1980) 263–93.
—— 'New Work for a Theory of Universals', *Australasian Journal of Philosophy*, 61/4 (1983), 343–77.
MACKIE, JOHN L., 'Causes and Conditions', *American Philosophical Quarterly*, 2/4 (1965), 245–64.
—— 'Counterfactuals and Causal Laws', *Analytical Philosophy*[1], ed. R. J. Butler (Oxford, 1966), 65–80.
—— *Truth, Probability, and Paradox* (Oxford, 1973).
—— *The Cement of the Universe* (Oxford, 1974).
McTAGGART, J. M. E., *The Nature of Existence*, ii (Cambridge, 1927).
MEHLBERG, HENRYK, 'Physical Laws and Time's Arrow', *Current Issues in the Philosophy of Science*, ed. H. Feigl and G. Maxwell (New York, 1961), 105–38.
MELLOR, D. HUGH, *The Matter of Chance* (Cambridge, 1971).
—— (ed.), *Science, Belief, and Behaviour* (Cambridge, 1980).
—— 'Comment', *Explanation*, ed. S. Korner (Oxford, 1975), 146–52.
—— 'Necessities and Universals in Natural Laws', in *Science, Belief, and Behaviour*, ed. D. H. Mellor (Cambridge, 1980), 105–25.
PAP, ARTHUR, 'Nominalism, Empiricism and Universals: I', *Philosophical Quarterly*, 9 (1959), 330–340.
PEARS, D. F., 'The Priority of Causes', *Analysis*, 17 (1956–7), 54–63.
POPPER, KARL R., 'The Arrow of Time', *Nature*, 177 (1956), 538.
—— Note, *Nature*, 178 (1956), 382.
—— Note, *Nature*, 179 (1957), 1297.
—— Note, *Nature*, 181 (1958), 402–3.
—— *The Logic of Scientific Discovery* (London, 1961).
—— 'A Revised Definition of Natural Necessity', *British Journal for the Philosophy of Science*, 18 (1967), 316–21.
—— 'Suppes's Criticism of the Propensity Interpretation of Probability and Quantum Mechanics', *The Philosophy of Karl Popper*, ed. P. A. Schilpp (La Salle, Ill., 1974), 1125–39.
RAILTON, PETER, 'A Deductive–Nomological Model of Probabilistic Explanation', *Philosophy of Science*, 45 (1978), 206–26.

RAMSEY, FRANK PLUMPTON, *The Foundations of Mathematics*, ed. R. B. Braithwaite (Paterson, NJ, 1960). 'Universals', pp. 112–34; 'Theories', pp. 212–36; 'General Propositions and Causality', pp. 237–55.

REICHENBACH, HANS, *The Theory of Probability* (Berkeley, Calif., 1949).

—— *Nomological Statements and Admissible Operations* (Amsterdam, 1954).

—— *The Direction of Time* (Berkeley, Calif. 1956).

—— *The Philosophy of Space and Time* (New York, 1958). First published as Philosophie der Raum-Zeit-Lehre, Berlin, 1928.

—— *Laws, Modalities, and Counterfactuals*, ed. W. Salmon (Berkeley and Los Angeles, Calif., 1976).

RESCHER, NICHOLAS, *Hypothetical Reasoning* (Amsterdam, 1964).

ROBB, ALFRED A., *A Theory of Time and Space* (Cambridge, 1914).

—— *The Absolute Relations of Time and Space* (Cambridge, 1921).

RUSSELL, BERTRAND, *The Problems of Philosophy* (London, 1912).

SALMON, WESLEY, 'Verifiability and Logic', *Mind, Matter and Method*, ed. P. K. Feyerabend and G. Maxwell (Minneapolis, Minn., 1966), 354–76.

—— 'Theoretical Explanation', *Explanation*, ed. S. Korner (Oxford, 1975), 118–43.

—— *The Foundations of Scientific Inference* (Pittsburgh, 1976).

—— 'An "At-At" Theory of Causal Influence', *Philosophy of Science*, 44 (1977), 215–24.

—— 'Laws, Modalities, and Counterfactuals', *Synthese*, 35 (1977), 191–229.

—— 'Why Ask "Why?"?', *Proceedings and Addresses of the American Philosophical Association*, 51/6 (1978), 683–705.

SCHEFFLER, ISRAEL, *The Anatomy of Inquiry* (New York, 1963).

SCHILPP, PAUL ARTHUR (ed.), *The Philosophy of Rudolf Carnap* (La Salle, Ill., 1963).

—— (ed.), *The Philosophy of Karl Popper* (La Salle, Ill., 1974).

SCOTT, DANA, 'Existence and Description in Formal Logic', *Bertrand Russell: Philosopher of the Century*, ed. R. Schoenman, (London, 1967).

SCRIVEN, MICHAEL, 'Randomness and the Causal Order', *Analysis*, 17 (1956–7), 5–9.

SHOEMAKER, SYDNEY, 'Causality and Properties', *Time and Cause*, ed. P. van Inwagen (Dordrecht, 1980), 109–35.

SKYRMS, BRIAN, 'Resiliency, Propensities, and Causal Necessity', *Journal of Philosophy*, 74 (1977), 704–13.

—— *Causal Necessity* (New Haven and London, 1980).

SMART, J. J. C., 'Causal Theories of Time', *Basic Issues in the Philosophy of Time*, ed. E. Freeman and W. Sellars (La Salle, Ill., 1971), 61–77.

SMOLUCHOWSKI, M. 'Gültigkeitsgrenzen des Zweiten Hauptsatzes der Warmetheorie', *Œuvres*, 2 (1927), 361–98.

Sosa, Ernest (ed.), *Causation and Conditionals* (Oxford, 1975).

Stalnaker, Robert C. and Thomason, Richmond H., 'A Semantic Analysis of Conditional Logic', *Theoria*, 36 (1970), 23–42.

Suchting, W. A. 'Regularity and Law', *Boston Studies in the Philosophy of Science*, 14, ed. R. S. Cohen and M. W. Wartofsky (New York, 1974), 73–90.

Suppes, Patrick, *A Probabilistic Theory of Causality* (Amsterdam, 1970).

Swoyer, Chris, 'The Nature of Natural Laws', *Australasian Journal of Philosophy*, 60/3 (1982), 203–23.

Szczerba, L. and Tarski, Alfred, 'Metamathematical Properties of Some Affine Geometries', *Logic, Methodology, and Philosophy of Science*, ed. Y. Bar-Hillel (Amsterdam, 1965), 166–78.

Taylor, Richard, *Action and Purpose* (New Jersey, 1966).

Tooley, Michael, 'The Nature of Laws', *Canadian Journal of Philosophy*, 7/4 (1977), 667–98.

—— 'Laws and Causal Relations', *Midwest Studies in Philosophy*, 9, ed. P. A. French, T. E. Uehling, and H. K. Wettstein (Minneapolis, Minn., 1984), 93–112.

van Fraassen, Bas C., *The Scientific Image* (Oxford, 1980).

Wright, G. H. von, *Explanation and Understanding* Ithaca, (NY, 1971).

INDEX

350   *Index*